Jeff Osfoor
Mobile, AL.
March 17, 2016

IMAGES
of America
USS ALABAMA

The USS *Alabama*'s bell is now located at the entrance of the visitor center and ship's store at USS Alabama Battleship Memorial Park. Visitors to the park are welcome to ring the bell. (Courtesy of Kent Whitaker.)

ON THE COVER: Crew members are seen on board in this aerial photograph of the USS *Alabama* (BB-60) early on commissioning day, August 16, 1942, at the Norfolk Naval Shipyard. The number three turret is located aft. At the time, the turret, like much of the *Alabama*, was still under construction. The top armor plating had also yet to be installed. (Courtesy of USS Alabama Battleship Memorial Park.)

Kent Whitaker on behalf of the
USS Alabama Battleship Memorial Park

Foreword by Bill Tunnell

Copyright © 2013 by Kent Whitaker on behalf of the USS Alabama Battleship Memorial Park
ISBN 978-1-4671-1021-1

Published by Arcadia Publishing
Charleston, South Carolina

Printed in the United States of America

Library of Congress Control Number: 2013930916

For all general information, please contact Arcadia Publishing:
Telephone 843-853-2070
Fax 843-853-0044
E-mail sales@arcadiapublishing.com
For customer service and orders:
Toll-Free 1-888-313-2665

Visit us on the Internet at www.arcadiapublishing.com

This book is dedicated to the crew of the USS Alabama and to all who have served this country—past, present, and future.

Contents

Acknowledgments		6
Foreword		7
Introduction		8
1.	The Ships Named *Alabama*	11
2.	Commanding USS *Alabama*	27
3.	*Alabama*'s Armament and "Big Guns"	39
4.	Working Aboard *Alabama*	53
5.	Life on *Alabama*	67
6.	Faith and Healing at Sea	83
7.	Earning Nine Battle Stars	93
8.	Bringing *Alabama* Home	109
About the USS Alabama Battleship Memorial Park		127

Acknowledgments

Bill and I would like to acknowledge the people and organizations that made this book possible. It could not have been written without the service of the World War II crew members of the USS *Alabama* (BB-60). Over the years, they have volunteered time, knowledge, and memories in order to help make *Alabama*'s history as complete as possible.

We would also like to acknowledge the Battleship Commission, the volunteers and fundraisers who brought *Alabama* home, and the current staff and volunteers of USS Alabama Battleship Memorial Park. They have all worked over the years to preserve her story.

Bill and I would also like to acknowledge Shea McLean, the curator of USS Alabama Battleship Memorial Park, and his staff—Todd Kreamer, Brittany Comiskey, Nora Ives, and Chip Dobson—for their tireless work in telling *Alabama*'s story.

Unless otherwise noted, all photographs in this book are courtesy of the USS *Alabama* and Battleship Memorial Park.

On a personal note, Bill and I would like to thank our wives, Mary Allyson Nagem Whitaker and Cynthia Gregory Tunnell, as well as our families. Thanks for letting us take the time to tell the story of the ship that means so much to both of us. Lastly, Macee Whitaker: thanks for your service to your country and for riding shotgun with me on all of the military history trips we have taken.

Foreword

Welcome to USS Alabama Battleship Memorial Park!

Dedicated to all Americans who have worn the uniform of every branch of the United States Armed Forces, the self-supporting Battleship Memorial Park is anchored by the World War II heroine battleship USS *Alabama* (BB-60). Although an agency of the state of Alabama, the park has never received any city, county, state, or federal funds for daily operations since its opening in 1965.

The park has been open to the public since January 9, 1965, exactly 18 years from the date of her decommissioning in 1947 in Bremerton, Washington. Some 14 million visitors have graced the decks of the *Alabama* since that date. The statewide economic impact of her being in Mobile has meant more than $500 million just from admission revenues alone!

Battleship *Alabama* is the "star" of the park, although other military artifacts are displayed. They include the submarine USS *Drum* (SS-228), now America's oldest existing submarine, and, along with Battleship *Alabama*, a National Historic Landmark. A total of 28 historic aircraft from the past 80 years are also displayed in the park, along with tanks, memorials, and other artifacts.

Everything about the battleship is big: it is longer than two football fields, had a crew of 2,500 men, and weighs about 90 million pounds or so. Even the park reflects the size of the ship, as 2.9 million cubic yards of Mobile Bay bottom were dredged to create the first 75 acres of the now 155-acre park. And oh yes, her keel is 24 feet below the waters of Mobile Bay.

Author Kent Whitaker has had a relationship with the USS *Alabama* since his infancy. He will take you on a spellbinding tour from her 1940 inception, through the fiercest of action during World War II, and continue following her trek after "retirement" to Mobile, where she stands today as a silent tribute to the Greatest Generation, her guns never again to fire in anger.

So please enjoy these images and words as the USS *Alabama* comes alive in your hands. And, yes, before I forget, come see her for yourself!

—Bill Tunnell, Executive Director

INTRODUCTION

The USS *Alabama* (BB-60) is a World War II South Dakota–class battleship. She is one of several American military vessels named after the state of Alabama.

USS *Alabama* and her three sister ships were the second series of battleships named South Dakota class. After World War I, several countries, including the United States, signed treaties that limited the capacity of warships. A 1922 treaty rendered the original South Dakota–class battleships already under construction obsolete, as they were over the limits set in the agreement. Construction was stopped, and the original ships were scrapped by 1923. The scrapped ships were to be named USS *South Dakota* (BB-49), USS *Indiana* (BB-50), USS *Montana* (BB-51), USS *North Carolina* (BB-52), USS *Iowa* (BB-53), and USS *Massachusetts* (BB-54).

In the 1930s, military leaders realized that there would be a need for updated "fast" battleships, aircraft carriers, and other naval vessels. A new war was engulfing much of the world. The scrapping of the original South Dakota–class ships in 1923 allowed naval planners the opportunity to modernize. The South Dakota–class battleship was brought back to the drawing board, redesigned, and funded. The new line of South Dakota–class ships borrowed some of the names from the ships scrapped in 1923. The new ships were named the USS *South Dakota* (BB-57), The USS *Indiana* (BB-58), the USS *Massachusetts* (BB-59), and the USS *Alabama* (BB-60).

The new battleships were more compact and featured nine 16-inch "Big Guns" in three turrets. This was three fewer "Big Guns" and one less turret than the original South Dakotas. They also had improved armor and protection below the waterline, and they were faster. All four of the new South Dakota–class battleships were under construction before the United States entered World War II in 1941 after the bombing of Pearl Harbor. Construction on the *South Dakota*, *Indiana*, and *Massachusetts* all began in 1939. The keel of *Alabama*, the last of the four, was laid down on February 1, 1940.

The American people were united behind the cause of defeating the Axis powers. Factories around the country switched from making only consumer items to also producing a wide range of materials for the war effort. The same effort to supply the troops took place in American shipyards. The crew building the *Alabama* sped up work on the country's newest battleship. More than 3,000 men and women, working 24 hours a day for 30 months, finished the $80 million project nine months ahead of schedule. The USS *Alabama* (BB-60) was christened on February 16, 1942, by Henrietta Hill, the wife of Sen. Lister Hill. *Alabama* was commissioned on Sunday, August 16, 1942.

Alabama displaced 35,000 tons when she was completed. When fully loaded, she weighed 45,000 tons, or 90 million pounds. She is 680 feet long and 108 feet and 2 inches abeam (wide). She is 194 feet tall from her keel (bottom) to top light. *Alabama* could steam at 28 knots, or almost 32 miles per hour. The *Alabama* and the other South Dakota–class battleships were also intended to serve as flagships.

South Dakota–class battleships were designed for providing shore bombardment and antiaircraft defense for aircraft carriers. *Alabama* carries a total of 129 guns. Each of the three large turrets house three 16-inch, 45-caliber guns, for a total of nine "Big Guns." Each of the nine could propel a 2,700-pound projectile more than 20 miles with great accuracy. The three large turrets were protected by 18 inches of armor.

The armament of the ship also includes 10 smaller side turrets. Each of the side turrets house two five-inch, 38-caliber guns. There are also 12 installations of four 40-millimeter guns. *Alabama* was originally fitted with 52 cannons, each 20 millimeters, which complemented the antiaircraft battery during close encounters from enemy warplanes.

USS *Alabama* underwent sea trials in the Chesapeake Bay under the command of Capt. George B. Wilson. She became known by her 2,500-man crew as the "Mighty A." When the sea trials were over, *Alabama* was assigned to the Atlantic theater of operations, and joined the war effort in early 1943 under Capt. Fred D. Kirkland. She was assigned to the British Home Fleet in the North Atlantic. Her Atlantic missions included the Murmansk Run, in which she protected British and Russian convoys delivering supplies to Russia. Years later, *Alabama* became the only American ship honored by the former Soviet Union for its role in helping to protect the Russian fleet during World War II.

Allied military planners also hoped that the presence of the *Alabama*, the newest battleship in the US Navy, in the North Atlantic would draw the German dreadnought *Tirpitz* from its hiding. Leaders believed that if *Tirpitz* made an appearance in order to fight *Alabama*, then Allied forces would have a better chance of sinking her. *Tirpitz* did not take the bait, and was not sunk until several years later, by aerial attack.

In August 1943, the *Alabama* left the North Atlantic and headed to Norfolk, Virginia, where she was overhauled, repaired, painted, and updated. After leaving Norfolk, *Alabama* headed south along the coast and eventually passed through the Panama Canal. Her new mission was to serve in the South Pacific as part of the Pacific Third Fleet. Because of a tight schedule, one modification was left to be finished at sea by the crew: the installation of a new SK-2 radar system.

The role of *Alabama* was a versatile one as she served across the South Pacific. She provided everything from fire support and antiaircraft screening for fast carrier forces to shore bombardments during landings. *Alabama* served in Efate, New Hebrides; the Gilbert Islands; the Marshall Islands; Truk; Palau; Yap; Ulithi; the Marianas, the Hollandia landings; and the invasion and capture of Saipan in June 1944.

The work done to install the new SK-2 radar system by the *Alabama*'s crew while at sea proved to be vital. In 1944, the *Alabama* was with Task Group 58.7 when the Japanese navy launched an aerial attack against the Pacific Third Fleet in the Philippine Sea near the Marianas. *Alabama*'s new SK-2 radar detected the massive group of incoming enemy ships when they were 190 miles away and confirmed the approaching Japanese ships at 140 miles away. This allowed plenty of time for the Allied fleet to launch aircraft and ready defenses. The battle is now known as the "Great Marianas Turkey Shoot." In the lopsided Allied victory, due in part to the early warning provided by the *Alabama*'s new radar, Japanese forces lost close to 500 aircraft and most of their experienced pilots.

Capt. Vincent R. Murphy took command of the USS *Alabama* in August 1944. Under his command, *Alabama* saw continued action while taking part in continuing operations in the South Pacific. *Alabama* served at the Battle of the Philippine Sea and took part in the capture and occupation of Guam. She served during attacks on Japanese forces in Palau, Yap, Ulithi, and the western Caroline Islands. *Alabama* also saw heavy action in Okinawa, Luzon, and the Surigao Strait during the Battle of Leyte Gulf and the Battle of Cape Engaño. She fought during the liberation of the Philippines as a member of Task Force 38 and operated off Surigao Strait as part of "Enterprise screen." She later steamed north to challenge the Japanese Central Force.

In December 1944, *Alabama* and several other ships were caught in a typhoon. The storm was so violent that three American destroyers were lost. Rough seas caused *Alabama* to roll in excess of 30 degrees. After two years at sea and the survival of the typhoon, the "Mighty A" was in need of repairs. While serving in Ulithi on Christmas Eve, Captain Murphy announced that his crew steer *Alabama* stateside for badly needed repairs.

Alabama stopped at Pearl Harbor en route to the continental United States. She then proceeded to the Puget Sound Shipyard at Bremerton, Washington, where Capt. William B. Goggins took command on January 18, 1945. When repairs were finished, *Alabama* once again set a course for Pearl Harbor. She then moved on to Okinawa and provided protective firepower for landing Allied forces. *Alabama* then took part in the shelling of the islands of Japan. On August 15, 1945, Japan surrendered.

The USS *Alabama* (BB-60) had the honor to lead the Allied fleet into Tokyo Bay the day after the Japanese formally signed surrender documents, on September 2, 1945. Marines serving on *Alabama* landed in Japan as part of the Allied occupation force. *Alabama* then continued to serve by providing coastal patrols before making her last voyage across the Pacific. On September 20, 1945, the "Mighty A" set a course for the United States. She was now also known as the "Lucky A" because none of her crew had been lost to enemy fire. While en route to the states, she stopped at Okinawa to pick up 350 sailors and servicemen who needed a ride home.

Alabama arrived on the West Coast and made several stops at various ports as she ferried servicemen home. Eventually, she arrived at Bremerton, Washington, where she was decommissioned on January 9, 1947. Capt. Edward H. Pierce was her final commanding officer. *Alabama* was now part of the so-called "mothball fleet."

The Navy announced in 1962 that the decommissioned *Alabama* would be scrapped. Citizens of the state of Alabama responded by seeking permission to bring the famed ship to its namesake state. Schoolchildren across the state saved their pocket change and partnered with businesses to raise the money needed to bring the *Alabama* to her new home. It took almost three months to tow the 35,000-ton World War II battleship from Bremerton, Washington, through the Panama Canal to its final resting place in Mobile. It was the longest nonmilitary tow in history. She received a hero's welcome on September 14, 1964, when she arrived at the newly created USS Alabama Battleship Memorial Park. The boat opened to public tours on January 9, 1965.

The USS *Alabama* (BB-60) logged 218,000 miles during World War II. Her crews shot down 22 enemy aircraft and she earned nine battle stars. Today, she is the centerpiece of USS Alabama Battleship Memorial Park. The park honors all men and women who have served in the military and plays host to various patriotic events throughout the year. The park also features the World War II–era submarine USS *Drum*, and is home to several veterans' memorials. There is also a military aircraft museum featuring an A-12 Blackbird, fighter planes, and fighter jets, as well as several static pieces, including tanks, helicopters, and even a Civil War–era submarine replica. Both *Alabama* and *Drum* are National Historic Landmarks, and Battleship Memorial Park has hosted more than 14 million visitors to date.

One

THE SHIPS NAMED ALABAMA

The USS *Alabama* (BB-60) is not the first to bear the name of the 22nd state in the union. The name Alabama was first assigned to a ship whose keel was laid in 1819. If completed, she would have been a 74-gun "ship of the line." Work was slow until the outbreak of the Civil War, when she was completed and renamed the USS *New Hampshire*.

The first *Alabama* was a 56-ton revenue cutter acquired on June 22, 1819, that captured a dozen pirate and slave ships. She served in the Caribbean and the Gulf of Mexico in the 1820s. The second *Alabama* was a 700-ton side-wheel steamer transferred to the Navy on March 3, 1849. She carried troops during the Mexican-American War but was deemed unfit for use and sold in October 1849.

The third *Alabama* was a side-wheel steamer commissioned on September 30, 1861. Duties included transporting troops and cargo during the Civil War. She was decommissioned in July 1865.

The Confederate raider CSS *Alabama* captured or sank 69 Union vessels during the Civil War. She was built in Liverpool, England, in 1862 and never made port in the states. She was sunk by the USS *Kearsarge* near France in 1864.

The USS *Alabama* (BB-8) was commissioned on October 16, 1900. She was the flagship for Division I, Battleship Force, the Atlantic fleet during World War I. She was decommissioned on September 15, 1921. She was used as a target by the War Department during the testing of aircraft attempting to sink ships by bombing. She was sunk on September 27, 1921.

A second *Alabama* may have served during World War I. A 69-foot motorboat named *Alabama* was inspected by the Navy in 1917 and placed on the rolls of the Naval Coast Defense Reserve. If called to service, her designation would have been USS *Alabama* (SP-1052).

The USS *Alabama* (SSBN-731) is a fleet ballistic missile submarine commissioned in 1985. Her conversion to a Trident D5 was completed in February 2009.

This engraving by Louis Le Breton shows CSS Alabama sinking after engaging in battle with the USS Kearsarge off the coast of France on June 19, 1864. Both Alabama and Kearsarge had their names live on as World War I–era US Navy battleships.

The USS Kearsarge was commanded by Capt. John A. Winslow. The CSS Alabama was commanded by Capt. Raphael Semmes. After the war ended, Captain Semmes settled in Mobile, where he practiced law. His great-grandson served on USS Alabama (BB-60) during World War II. This Civil War–era image shows Captain Semmes aboard CSS Alabama.

This photograph shows Illinois-class battleship USS *Alabama* (BB-8) in 1904. *Alabama* was larger and faster than the USS *Kearsarge* (BB-5), which was designed for coastal defense. Both *Alabama* and *Kearsarge* were laid on December 1, 1896. Both ships were named after Civil War–era ships. The *Alabama*'s original masts are seen here before the boat was placed in reserve and decommissioned in 1909.

This photograph shows crew members of the USS *Alabama* (BB-8) having a little fun while demonstrating the size of her guns. A crew member is lying down inside one of the *Alabama*'s 13-inch "Big Guns." *Alabama* (BB-8) had four 13-inch guns in two turrets.

Alabama (BB-8) underwent extensive modernization from late 1909 to early 1912. The most striking difference was the modification of her original masts to cage-style masts. Also notable is her new paint job, which replaced her original white paint. Most of her World War I service consisted of reserve, training, and fleet service along the coast of the United States and Central America.

USS *Alabama* (BB-8) was fully decommissioned in 1920 and transferred to the War Department in 1921, where she served as a bombing target for the Army Air Service. This photograph, taken on September 23, 1921, shows a twin-engine Army Martin bomber flown by legendary airman Billy Mitchell. It was taken during a test to show Navy leaders that a plane flown off a ship could sink a battleship.

During the laying of the keel ceremony for the USS *Alabama*, representatives from the Army and Navy joined representatives from the state of Alabama. These two photographs show the ceremony underway for the driving of ceremonial rivets at the Norfolk Navy yard on February 1, 1940. Representatives of the military are seen above and local representatives from the state of Alabama are below.

This photograph of the keel-laying ceremony for the USS *Alabama* shows the size of the project. The main stage area is center right, with one person kneeling. The whole event took place on temporary wooden walkways, which were dismantled and reused as needed as the building project progressed. Material could be reused, burned, or scrapped. At the time, efforts were in place to try to reduce waste as much as possible.

The USS *Alabama* (BB-60) is seen here prior to its christening. Visible is the platform on which the participants stood during the event (bottom left). Also visible on deck are the housings for turret one and turret two. Both housings are covered with a dome-like cap to protect the ship's interior from the elements until the gun turrets are in place.

The photograph above shows *Alabama*'s launching schedule board. This board kept track and gave the exact order in which various blocks, frames, wedges, and other things holding *Alabama* in place were to be removed.

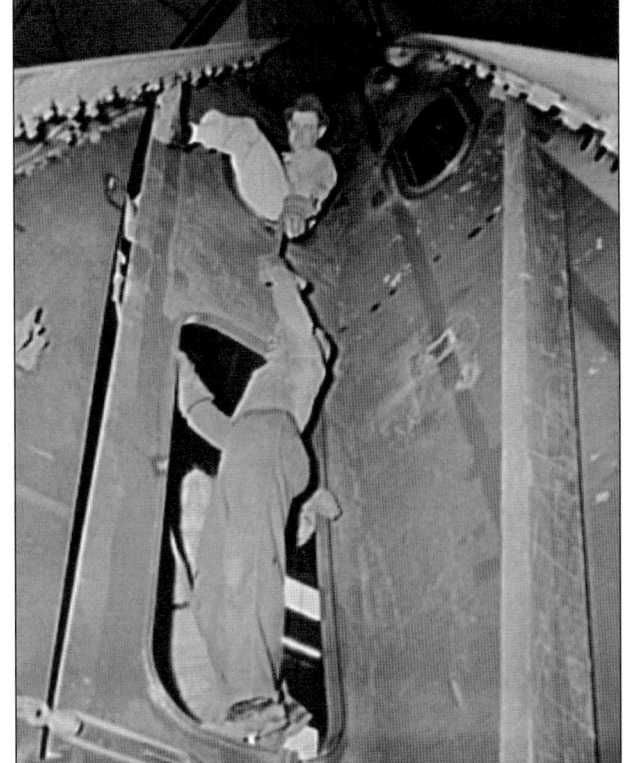

Pipefitters are seen here working on the interior of *Alabama*. More than 3,000 men and women, working 24 hours a day for 30 months, finished the $80 million project nine months ahead of schedule.

Henrietta McCormick Hill (center), the sponsor of *Alabama* and the wife of Sen. Lister Hill of Alabama, waits to christen the ship. In her book *A Senator's Wife Remembers*, she described the day as being a celebratory one despite the light rain.

Henrietta McCormick Hill christens the *Alabama* on February 16, 1942. She later said that the ship moved before she was ready. Having been told that it would be bad luck if the bottle did not break, she swung as hard as possible. The remains of the bottle and the wire safety container that surrounded it were later presented to her in a gift box now displayed aboard *Alabama*'s main deck.

Joseph Lister Hill, seen here in his World War I uniform, represented the state of Alabama as a representative and a senator for more than 45 years. He was also a World War I veteran. Appropriations to build the South Dakota–class ships, including *Alabama*, were passed during his tenure. (Courtesy of Henrietta Hill Hubbard.)

As sponsor of the USS *Alabama*, Henrietta Hill was presented a set of silver (below). The workers in the shipyard raised money to buy the silver by donating pennies and nickels. Hill stated in her book *A Senator's Wife Remembers*, "I shall always treasure this beautiful gift since it comes to me from everyone connected with the construction of the *Alabama*." (Courtesy of Kent Whitaker.)

Alabama slips into the water as the dignitaries gathered on the christening stand watch. Scaffolding on top of *Alabama* gave several shipyard workers a bird's-eye view of the event. She was the largest ship ever built in the Norfolk Navy Yard, in Portsmouth, Virginia. The wave created when she hit the water splashed spectators on the other side of the river.

Alabama continues to slide into the water after being christened on February 16, 1942. The large openings at the top of the ship are for her anchors, which are not in place. At this point, the new ship has no way of controlling her path or progress.

Alabama is seen in these two photographs after she has left her dry dock. Above, tugboats move into place to contain her movement. Below, the tugs guide her to her new moorings in the shipyard to be fitted out. Workers and materials can be seen on her deck despite the mist and fog.

After being moved into position to be fitted out, *Alabama* is met by an old "namesake" friend. The crane ship to the left is the former battleship USS *Kearsarge*, which had been renamed and refitted as a crane ship. Extra ballast was added to her sides for better balance while lifting heavy loads. The size difference between a World War I battleship and the new *Alabama* is clearly visible.

This photograph shows the former USS *Kearsarge* lifting armor plating for turret number three. The ship whose namesake once sunk an *Alabama* was renamed and converted to a crane ship, but the name *Kearsarge* is still proudly painted on her hull. The name *Kearsarge* was assigned to the USS *Kearsarge* (CV-33), which was laid down in 1944 and commissioned in 1946.

These two photographs of *Alabama* show crew members beginning to gather in their dress white uniforms early on the day of her commissioning, August 16, 1942. The photograph above shows turrets one and two looking aft while the photograph below shows turret three looking forward towards the stern. The crane of the *Kearsarge* can be seen to the right below.

This photograph from the commissioning ceremony shows the aft portion of *Alabama*. The crane is not for construction; it is for lifting and moving the two airplanes that would be stationed on the ship. The flat surface on the left side of the image is one of the catapult systems used to launch her planes.

This picture, also taken during the commissioning, shows that work on *Alabama* was ongoing. Workers were still finishing items while Navy crews began preparing the ship for service. Armor has yet to be installed on top of the turret in this image.

During *Alabama*'s early shakedown cruises, members of the press were guests of the ship. Seen here from left to right are Sandor S. Klein of United Press, William Pickens of Transradio Press Service, Halilton Faron of the Associated Press, and Joseph Bars of the International News Service. The two men in the back left are unidentified.

In 1941, Abbott and Costello starred in the movie *In the Navy*, which was set on a fictitious USS *Alabama* while the real *Alabama* was still under construction. Interior shots were filmed on a soundstage while exterior shots were filmed on Navy bases. This Universal Pictures publicity still features Abbott (right) and Costello (center) with Shemp Howard (left) of the Three Stooges. (Courtesy of Kent Whitaker.)

Two

COMMANDING USS ALABAMA

The USS *Alabama* (BB-60) was commanded by eight men. Capt. George B. Wilson served as the first captain and led her through sea trails. Captain Wilson served from August 16, 1942, to March 17, 1943. When he suddenly fell ill, he was transferred off the battleship and Comdr. Neil K. Dietrich, who was already serving on board, took command for three days until a new captain arrived.

The next full-time captain was Capt. Fred D. Kirtland, who served from March 20, 1943, to August 12, 1944. Following Captain Kirtland was Capt. Vincent R. Murphy, who commanded *Alabama* from August 12, 1944, to January 18, 1945.

Capt. William B. Goggins was the next captain of the "Mighty A," from January 18, 1945, until November 26, 1945, when Capt. Edward H. Pierce took command, running the ship until August 12, 1946. *Alabama* had two more captains during the last days of her service: Capt. Claude F. Bailey, from August 12, 1946, to August 20, 1946, and Charles B. Phillips, from August 20, 1946, to January 9, 1947. Several admirals also visited *Alabama* while she served.

Capt. George B. Wilson became the first captain of the USS *Alabama* (BB-60), commanding her during early shakedown cruises. Captain Wilson was later promoted to rear admiral, and his World War I and World War II military decorations include the Mexican Service Medal, Victory Medal World War I, Grand Fleet Clasp, the Yangtze Service Medal, American Campaign Medal, European Campaign Medal, Bronze Star, and Victory Medal World War II.

These photographs show Capt. George B. Wilson and his officers posing for official photographs, with covers on (above) and covers off (below). They are, from left to right (seated) Comdr. D.F. Zimmerman, Comdr. Neil K. Dietrich (executive officer), Captain Wilson (center), Comdr. W.F. Karbach, and Comdr. Robert L. Campbell; (standing) Lt. Comdr. G.S. Everett, Lt. Comdr. K.V. Dawson, Comdr. E.J. Taylor, Comdr. L.E. Coley, and Lt. Comdr. George L. Markle (chaplain).

Captain Wilson (left) shakes hands with one of his crew members during an inspection. Veterans of *Alabama*'s first crew sometimes referred to Captain Wilson as "Sandbar" because once during an early shakedown cruise, when *Alabama* ran aground on an unmarked sandbar, Wilson ordered the crew to run from one side of the ship to the other in an effort to free her. She was eventually towed free.

Nothing was left unchecked by Captain Wilson, seen here making his rounds through the galley areas of *Alabama* during an inspection. These large cauldrons could be used for a variety of foods, ranging from soups and stews to beans, potatoes, and hot breakfast cereals.

Comdr. Neil K. Dietrich served as *Alabama*'s first executive officer and was a "plank holder," a crew member present at the time of a ship's commissioning. While not an official captain, Commander Dietrich was in charge of *Alabama* from March 17, 1943, to March 20, 1943, after Captain Wilson fell ill. Commander Dietrich left *Alabama* in September 1943 and retired in 1958 with the rank of rear admiral.

Capt. Fred D. Kirtland took command of USS *Alabama* (BB-60) on March 20, 1943, taking her to serve in the Atlantic theater of operations with the British Home Fleet. *Alabama* was based in the Orkney Islands, off the northern coast of Scotland. Her main duty was to protect convoys on the Murmansk Run, in the Arctic Ocean between Iceland, the British Isles, and the Russian coast.

Captain Kirtland is seen below when *Alabama* was serving in the Atlantic. Aside from working with the British Home Fleet, Allied military planners also hoped that the presence of *Alabama* would draw the German battleship *Tirpitz* out from hiding and into the open.

This photograph was taken on March 17, 1944, as Captain Kirtland (center) makes his way aft towards one of *Alabama*'s Kingfisher scout planes. The crew of *Alabama* stands at attention in order to say goodbye as Captain Kirtland prepares to disembark. Capt. Vincent R. Murphy soon took command of the "Mighty A."

Captain Kirtland climbs aboard one of *Alabama*'s Kingfisher scout planes as he departs the ship. Kirtland captained *Alabama* in both the Atlantic and Pacific. He was awarded the Legion of Merit "for exceptionally meritorious conduct as Commanding Officer," as well as a Commendation Ribbon "for distinguished service." Kirtland retired with the rank of vice admiral.

Capt. Vincent R. Murphy captained *Alabama* from August 12, 1944, to November 14, 1944. His service included the attacks leading to the capture of the southern Carolina Islands and the Palaus, as well as the attacks on Japanese military installations in the Philippines and more. Captain Murphy also served in World War I, commanded submarines between the wars, and was the navigator on the USS *Texas* (BB-35).

While on the staff of the commander-in-chief of the Pacific fleet in Pearl Harbor, Captain Murphy relayed the message, "Japanese air attack on Pearl Harbor, this is no drill" on December 7, 1941. Years after his 1946 retirement, he became executive vice president of the Navy Relief Society, serving there until 1962. Captain Murphy's war honors include the Legion of Merit, the Bronze Star, and the Navy and Marine Corps Medal.

On January 18, 1945, Capt. William B. Goggins assumed command of *Alabama* during her overhaul at Puget Sound Navy Yard in Bremerton, Washington. He then returned *Alabama* to Pacific operations and joined the Third and Fifth Pacific Fleets in May 1945. Captain Goggins commanded *Alabama* through many Pacific campaigns, including the bombardment of the Japanese home islands.

Captain Goggins left *Alabama* on November 26, 1945. His long list of honors includes the Purple Heart, Navy Commendation Ribbon, Legion of Merit, Gold Star with Combat Distinguished Device "V" for meritorious service, World War I and II Victory Medals, Atlantic Fleet Clasp, American Defense Service Medal, Fleet Clasp, Asiatic-Pacific Campaign Medal, American Campaign Medal, Philippine Defense Ribbon with Bronze Star, Philippine Liberation Ribbon, and the Navy Occupation Service Medal, Asia Clasp.

This photograph, taken from *Alabama*'s original war diary, shows Rear Adm. E.W. Hanson, who selected the "Mighty A" as his flagship. Rear Admiral Hanson was the commander of Pacific Battleship Division 9 in July 1944. His division of ships worked with a task group formed around the carrier *Bunker Hill*. His ships, including *Alabama*, screened Allied carriers as they conducted attacks and landings on Guam.

Below, Rear Adm. Richard E. Byrd boards *Alabama*. This method was used to move all types of items between ships, including injured crew from smaller boats who needed access to *Alabama*'s larger medical facilities, food, ammunition, mail, and other supplies. The lines were also used to move fuel hoses back and forth for refueling. This photograph was taken on July 26, 1945, in the Pacific.

Rear Adm. Richard E. Byrd, the famed explorer, was called back to service during World War II. He served in the Pacific and briefly visited operations in Europe. In July 1945, Rear Admiral Byrd was on board *Alabama* to observe the night bombardment of major industrial plants northeast of Tokyo. He was also present during the Japanese surrender.

Vice Adm. Willis A. Lee (left) is seen here on *Alabama* during a medal ceremony. Lee was promoted to vice admiral in 1944 as the commander of battleships for the Pacific fleet. In 1945, he was sent to the Atlantic to command a unit researching defenses against Japanese kamikaze planes. Vice Adm. Willis A. Lee died from a heart attack while serving in that position on August 25, 1945.

Rear Adm. John Franlin Shafroth (later promoted to vice admiral) is shown (right) after presentation of the Legion of Merit to an unidentified *Alabama* officer. Rear Admiral Shafroth reported aboard on August 12, 1945, as Commander Battle Squadron Two and Battleship Division Eight during the final days of the bombardment of Japan. Prior to World War II, he commanded the USS *Indianapolis*. Arguably the largest man in the Navy, his father was a U.S. senator and governor of Colorado.

Three

ALABAMA'S ARMAMENT AND "BIG GUNS"

The USS *Alabama* (BB-60) was called a battleship for a reason. World War II battleships were direct descendants of the military "ships of the line," which were often associated with pirate movies. That concept evolved into modern battleships that could deliver a blow while hopefully being farther away from an enemy attack.

The designers of the South Dakota class of battleships took this concept of packing a punch to heart. They were constrained on the size and number of "Big Guns" that a battleship could have because of treaties signed after World War I, so designers packed as much power onto the decks of the four new South Dakota–class battleships as they could. The USS *Alabama* and her three sister battleships carried an impressive armament.

Alabama had three main turrets that housed nine massive 16-inch guns, commonly called her "Big Guns." The main guns could launch projectiles that weighed as much as a small car more than 20 miles with amazing accuracy. There were also 10 smaller mounts in the secondary battery housing 20 five-inch guns, which packed a deadly blow. Alabama also carried two types of antiaircraft guns, for a total of 129 guns in her arsenal.

As if that was not enough, *Alabama* also carried a full detachment of Marines and all of their weapons and gear.

This 1945 photograph of *Alabama* provides a good visual of her total armament. Turrets one and two are visible showing six of her nine 16-inch "Big Guns." Also visible are her 20 twin-mount, five-inch, 38-caliber guns in 10 smaller turrets, as well as many of her 48 antiaircraft machine guns and 52 single antiaircraft guns.

These two photographs show crew members manning some of *Alabama*'s single-barrel, 20-millimeter antiaircraft guns. There were 52 of these guns on *Alabama*, and they were capable of firing quarter-pound shells at a rate of 450 rounds per minute.

This photograph shows the view from a 20-millimeter gunner's position. Gunners used their training and the large sight, seen here, to judge where they aimed and fired their weapons.

The photograph below shows the back of one of *Alabama*'s twelve 40-millimeter antiaircraft quad gun mounts. Each mount carried four guns, for a total of 48. Each gun could fire a two-pound shell at a rate of 160 rounds per minute. Also seen are several emergency life rafts, which were placed liberally around the deck.

Major league pitcher Robert William Andrew "Bob" Feller served aboard *Alabama* with one of the 40-millimeter antiaircraft gun crews. He was the first professional athlete to enlist after the attack on Pearl Harbor. Feller was a chief petty officer with six campaign ribbons and eight battle stars at the war's end. He served in both the North Atlantic and in the Pacific.

Bob Feller (center, facing camera) is seen here with his other 40-millimeter machine gun crew members during a training exercise aboard *Alabama*. While Feller went to great strides to just be a normal crew member, it was quite common to see him practice pitching on deck during downtime.

This photograph from the original USS *Alabama* war diary is simply captioned, "Machine Guns." It is a great example of how many men needed to cram into a small space in order to operate their assigned weapons.

Alabama's 10 smaller turrets contained her secondary battery. Each turret held two five-inch, 38-caliber guns, for a total of 20. A properly trained crew could fire 15 rounds per minute per gun. They could fire antiaircraft rounds, semi-fixed rounds, and some special-purpose illuminating rounds. The photograph below shows a smoke ring during live fire.

The main battery of any battleship is often nicknamed the "Big Guns." *Alabama* carried three turrets with three "Big Guns" each, for a total of nine. Each contains 16-inch, 45-caliber MK VI guns. Turrets one and two are located forward, with turret three aft of *Alabama*'s superstructure.

Turret one (forward) and turret two are seen here taking aim to the starboard side while *Alabama* navigates the North Atlantic in December 1942. Snow can be seen atop turret one. Each turret crew could fire each gun at a rate of two rounds per minute. In addition, each gun could be elevated and trained together or individually.

Alabama's "Big Guns" packed a punch. This photograph shows several things. First, the guns are working independently of each other, while some movement is from designed recoil. Secondly, the protective skirts, or bloomers, are clearly moving and puffing from the force of the blast. Also visible is the shockwave across the surface of the water.

Alabama is seen here as she fires a broadside in the Atlantic in 1943. The force of the blast is visible by the shockwave on the surface of the water. Also visible is the slight listing of *Alabama* in the water during the blast. Each round fired meant that more than 3,200 pounds of weight was leaving the ship.

Alabama's "Big Guns" also provided perfect platforms from which to watch. This crew member is on watch atop turret two during a heavy snow in the Atlantic. Also visible in the snow, on the center ladder, is one of *Alabama*'s 40-millimeter gun mounts.

The photograph of *Alabama* above was taken while she fired her 16-inch "Big Guns" in the Atlantic. The blast of the guns caused the snow on her turrets to turn into powder. Also visible is her "Measure 12" camouflage, on the sides of turrets one and two.

The caption for this photograph from the original *Alabama* war diary reads, "The *Alabama* Fires a Perfect Offset Straddle at the *Indiana*." It is a good example of how battleships could fight aggressively and defend themselves from both sides at the same time.

According to this photograph caption, it shows *Alabama*'s G.W. Ryder (left), seaman second class, as he works the scuttle inside of one of *Alabama*'s three turrets. Powder man F.L. McCain lifts one of the six powder bags needed to fire one projectile. Combined, the six silk-wrapped bags of powder weighed 540 pounds.

The South Dakota–class battleship's 16-inch guns fired projectiles that were as tall as a crew member. This photograph shows crew members R.C. Carlson (left) and E.M. Vickers on *Alabama*'s turret one using a hoist to lift one of the projectiles, which weighed up to 2,700 pounds each.

Here, crew members Carlson (left) and Vickers have lifted a projectile into a lift to send it to its next stop to be loaded. The projectile is as tall as the crew members. There were several projectiles used by *Alabama*, including armor piercing, target, high capacity, tracer, timed fuse, and primer.

Each of the six bags of powder needed to fire a projectile weighing nearly 100 pounds. The bags were made from silk, which allowed them to burn completely when fired. The bags were also color-coded. Six bags were used because one large bag would have been too heavy to handle.

Efficiency was important in every step of the firing of the "Big Guns." In order to save time and lessen tripping hazards, gunners secured towels to their arms. The gunner could visually check the breechblock of the 16-inch gun after it was fired while wiping down the "mushroom" at the same time. This photograph was taken on February 18, 1943.

Four

Working Aboard Alabama

The workday for the crew members of the USS *Alabama* (BB-60) never really stopped. While underway, the ship was alive with activity 24 hours a day. The majority of the crew worked normal hours, most of the time, but schedules were dependent on the situation. Things could change in an instant, depending on if the ship was docked, anchored, underway, or at general quarters.

Most often, the crew worked various shifts depending on their jobs and assignments. Many work hours took place during the course of a "normal" day. The first crewmen up were the bakers and cooks, who started baking bread and getting meals ready for the crew as early as 4:30 a.m.

Since *Alabama* was a battleship, its size allowed for more than normal work hours based on her armament. The ship had a large medical facility, a post office, a seamstress shop, machine shops, and several galley areas.

Aside from the normal activities involved with working on a battleship, the crew also spent much of their time training, cleaning, repairing, and keeping their vessel ready for combat. The good news was that a hot cup of coffee was available 24 hours a day for those on duty overnight.

This photograph shows a supply transfer between *Alabama* and *South Dakota* in 1943. Items ranging from basic supplies, movies, mail, food, medicine, and fuel could be transferred between ships on the move. If it could be "rigged," it could be sent between ships. Transfers of this type were common and often included chaplains being sent to smaller ships.

Members of *Alabama*'s crew perform a "sounding." Even though the ships were more modern than ever, sailors still used hand measurements to determine the depth of the water when in shallows. A weighted line was dropped into the water as measurements were relayed to the bridge. This was common every time a ship pulled into port.

Just because the boat was at sea did not mean an inspection could not be called. These two photographs show *Alabama*'s crew under inspection while in winter gear in the Atlantic (above) and in "whites" while serving in the Pacific (below).

Alabama's Navy crew members were not the only ones subject to inspection. Here, a detachment of Marines that served aboard *Alabama* is seen during an on-deck inspection.

Daily life aboard *Alabama* took a turn on December 18, 1944, when she, along with several other ships, rode out a typhoon. *Alabama* listed 30 degrees on several occasions. Her planes were damaged beyond repair and three destroyers were lost in the storm. The "Mighty A" rode out a second typhoon in 1945, which caused only superficial damage.

Reports do not have to be boring. In one of his report entries, Comdr. Robert L. Campbell, seen here, of the USS *Alabama*, noted the condition of the enemy after an assault by Allied forces on the island of Truk in the Pacific, writing, "The First Team of the American Navy left Truk a smoking shambles."

Comdr. Neil Dietrich is seen here. Part of being in the chain of command on any ship was checking daily paperwork and logs. In the foreground of this image, the gun turrets of a wooden model of *Alabama* can be seen. Also interesting is the framed picture hanging above Commander Dietrich's head. Artwork of this type was typical on many vessels.

Alabama had its own printing press, as did many ships in the Navy. The press was used to produce flyers, ship newspapers, and give announcements for the crew, as well as official reports. One favorite printed item among crews were holiday galley menus.

Machinist Mate Bill Soper (far right) is pictured with a group of crew members. A handwritten note on the back of this photograph, from the archives of the USS *Alabama*, says that these men were "Damn good friends."

These two photographs show *Alabama*'s Central Radio Receiving Room in December 1942. The crew members at right are, from left to right, unidentified, Chief K.A. Trexler (standing), L.H. Bethune, R.E. Christensen, Al Barkan, Lou Brown, and Allan B. Hildebrand.

A first class fire control technician is seen here during *Alabama*'s December 1942 shakedown cruise. His job was to close the firing circuits on the MK-43 "Big Guns" gimballed gyro, which stabilized the gun as orders were sent to the turrets. The room also operated as the firing station for the plotting room.

Crew members of the 16-inch "Big Gun" main battery fire control team are seen below in the plotting room during *Alabama*'s shakedown cruise in December 1942. The workday in the plotting room involved a number of tasks, including checking target information.

Five of *Alabama*'s crewmen pose for a photograph on the deck in the Pacific. The handwritten description on the back identifies them as being some of *Alabama*'s radiomen.

Taking a break from the normal workday, Rear Adm. Edward W. Hanson and officers from *Alabama* posed for this photograph. They are, from left to right, (first row) R.P. White, F. Robinson Jr., Admiral Hanson, R.N. Atkinson, R.S. Newman, and G.H. Finn; (second row) D.K. Lemmers, H.B. Hurst, Comdr. D.C. Lyndon, G.H. Niles, E.W. Johnson, and Ward McNally.

Crew members work to release the Pelican hook in order to free one of the anchors on *Alabama*. This process required an incredible amount of skill and safety. World War II battleships, depending on their class, had anchors with a combined weight upwards of 30,000 pounds. Each link in the chain weighed approximately 150 pounds on a South Dakota–class battleship.

The photograph below was taken from the USS *Alabama* war diary and shows an unidentified seaman second class tending the aft steering wheel, as a first class seaman and a chief petty officer keep an eye on his work and the gauges.

A workday while serving in the North Atlantic often included removing ice and snow from anything topside. *Alabama*'s crew is seen in these two photographs removing ice and snow from the "Big Gun's" bloomers as well as from the deck. Besides being a slipping hazard, extra snow and ice added unwanted weight to the battleship and could hamper equipment from working properly.

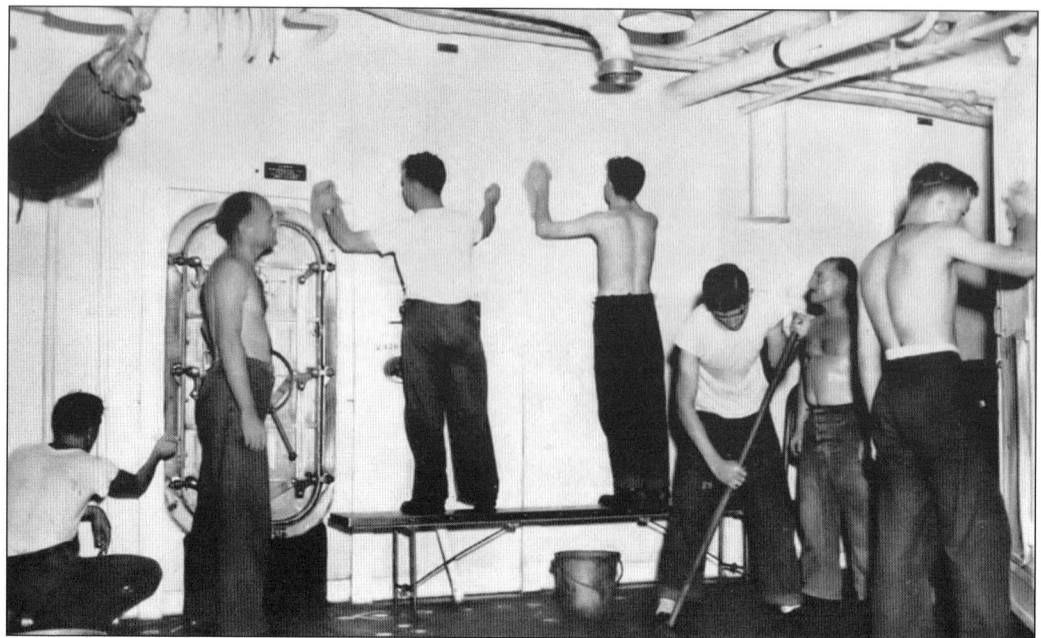

Some workdays aboard *Alabama* were "field days," when general cleaning and maintenance were performed. The days served several purposes, including keeping equipment clean and safe, identifying any repairs that needed to be made, and keeping crews busy during any downtime. Because they were at sea, many ships required constant cleaning due to the salty ocean air and the effects of condensation.

A chief petty officer checks and records the salinity content of freshwater that has been distilled from seawater by one of *Alabama*'s two large evaporators in the forward auxiliary machinery room.

A fireman aboard *Alabama* works a burner on one of the ship's boilers during her shakedown cruise in January 1943. Four boilers were used when underway, each operating at a steam pressure of 600 pounds per square inch. Others were brought online as needed, or when one was taken offline for maintenance or repair.

Regular workdays often included training. These *Alabama* crew members are training on a practice-loading mechanism for a five-inch, 38-caliber gun on the open deck. Any Navy combat vessel that carried this type of gun also carried a practice-loading machine. *Alabama* carried 20 of the guns, mounted in 10 mounts.

A seamstress shop was located aboard *Alabama* and other larger naval vessels. The shop serviced her own crew, and, if time allowed, sometimes the crews of smaller vessels without the luxury of their own shops. The shop workers could sew and repair uniforms, linens, and even tarps.

Combat photographers in the military served in all service branches and in all theaters of the war. Without their dedication and service, no World War II military book would be possible. Here, an unidentified combat photographer serves aboard *Alabama*.

Five
LIFE ON ALABAMA

Life on the USS *Alabama* was not always centered on being ready for battle. There were times when a crew member had a few moments off. The "Mighty A" served as a home away from home for her men. At sea, versions of everyday things were in place for her crew to take advantage of during downtimes.

Alabama's large deck made it possible for movie night to be an outdoor event almost every rainless evening. The same expansive deck also served as an area where large sporting events took place. Leisure time activities were not always large-scale events, however. Bob Feller and a few men could be found tossing a baseball around on occasion.

Events large and small served as a way for the men to form a bond with their shipmates. Pulling for the ship's baseball, basketball, or boxing team when they took on the best from another ship took the place of pulling for a hometown club.

Sailors could also enjoy mundane activities, like simply taking a moment for a haircut and a shave or enjoying an ice cream from the "gedunk." Sometimes, enjoying life on *Alabama* simply meant writing a letter home, attending a church service, or quietly enjoying a good book in the reading room.

Often, the best part of life on *Alabama* was being able to leave for liberty. This photograph, taken while *Alabama* served in the North Atlantic, shows crew members lined up for an orderly departure from the ship for some free time ashore.

This photograph shows *Alabama* as she makes her way through one of the locks on the Panama Canal. Since the ship was guided by lock workers, the crew took the opportunity to enjoy the sights and grab some fresh air. A determining factor in modern shipbuilding is that any ship has to be able to fit through the Panama Canal locks.

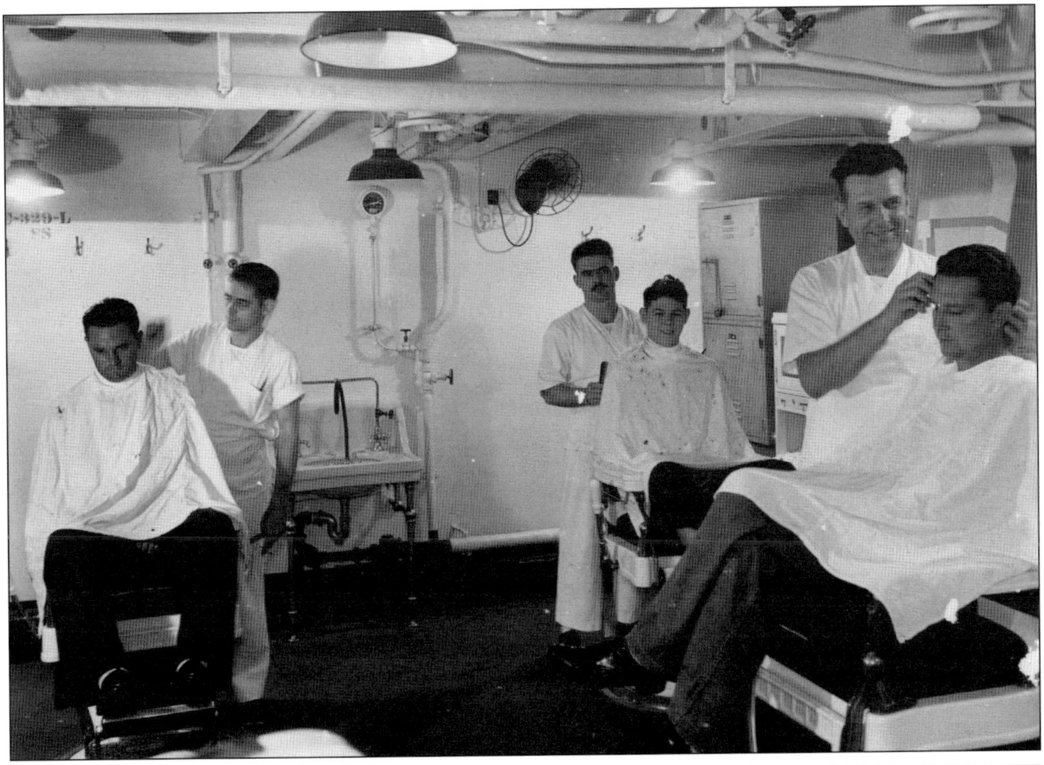

Above, several members of *Alabama*'s crew take a few minutes to get their hair trimmed in the ship's barbershop. The shop also served as the barbers' sleeping quarters. The mirrors in the shop were highly polished pieces of stainless steel, as were all mirrors on the ship, in order to prevent accidents. *Alabama*'s barbershop was manned by three barbers and three apprentice barbers.

Several members of *Alabama*'s crew sported full beards, as can be seen in this photograph of an unidentified crew member. The rules for facial hair varied from ship to ship and from captain to captain.

Berthing areas were located all over *Alabama*. The crew numbered 2,500 on average and could increase or decrease depending on the number of passengers she ferried. Many free spaces along the walls had fold-down bunks in order to save space. Hooks hung above head in many areas of the ship, to accommodate any number of temporary hammocks that might be needed. Helmets are at the ready in these photographs, in case general quarters is sounded.

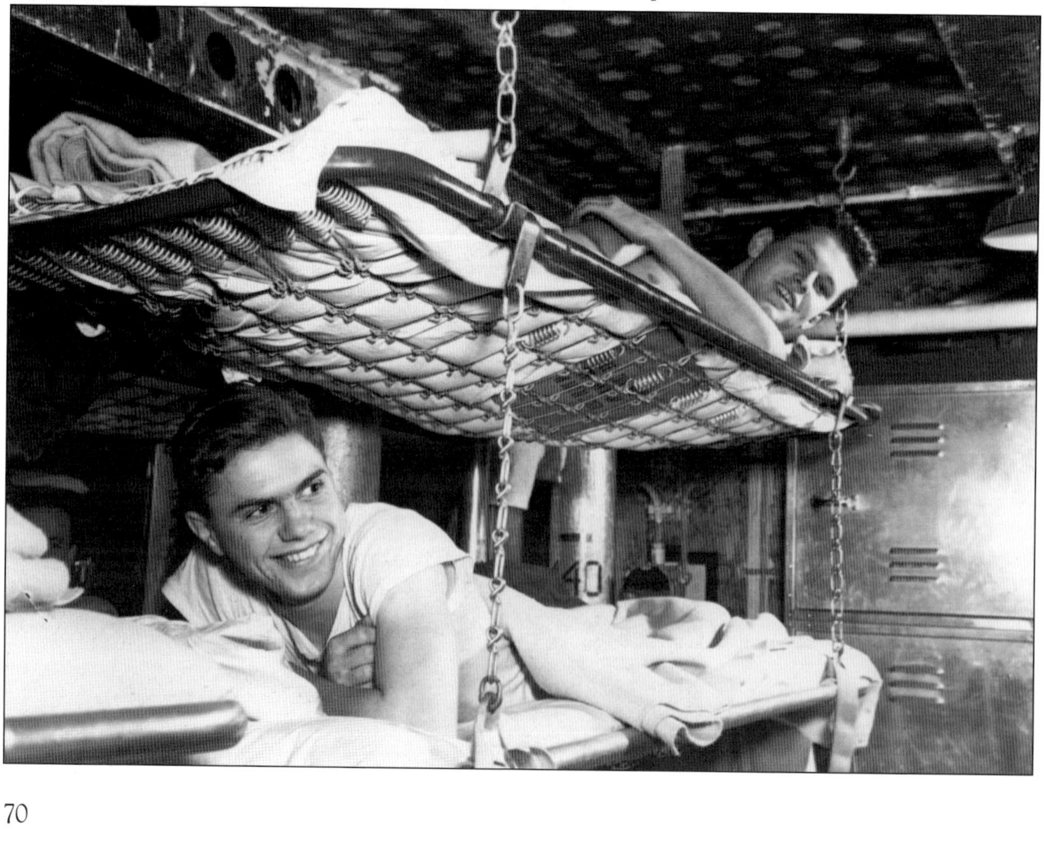

Bakers are seen here on *Alabama*. The bakers made enough bread for the 2,500-member crew on an almost daily basis. Above, loaves of bread are rising in the foreground in front of the men posing for a photograph. At right, bakers prepare dough, using a scale to properly measure ingredients.

Alabama's bakeshop crew produced 550 loaves of bread a day, in addition to cakes, pies, biscuits, and cornbread. Here, bread is stacked on end in order to save space. It was not uncommon for larger ships with bakeshops to offer smaller ships a few extra loaves if needed, or in trade.

Poultry is selected for use in *Alabama*'s butcher shop. The butcher shop was supplied with frozen meat every morning from *Alabama*'s freezers, called "reefers." The meat was thawed out and used that day. Nearby was the vegetable prep room, where all fruits and vegetables were washed and prepared. Potatoes were kept in the "spud locker."

The "gedunk" stand was *Alabama*'s ice cream and soda shop, which served snacks, malts, ice cream, sodas, and shakes for a minimal price. Coffee was also available. The slang term "gedunk" refers to the mechanical "gee-dunk" sound that vending machines made after an item was selected. Ice cream was made daily and was a favorite item to trade with other ships.

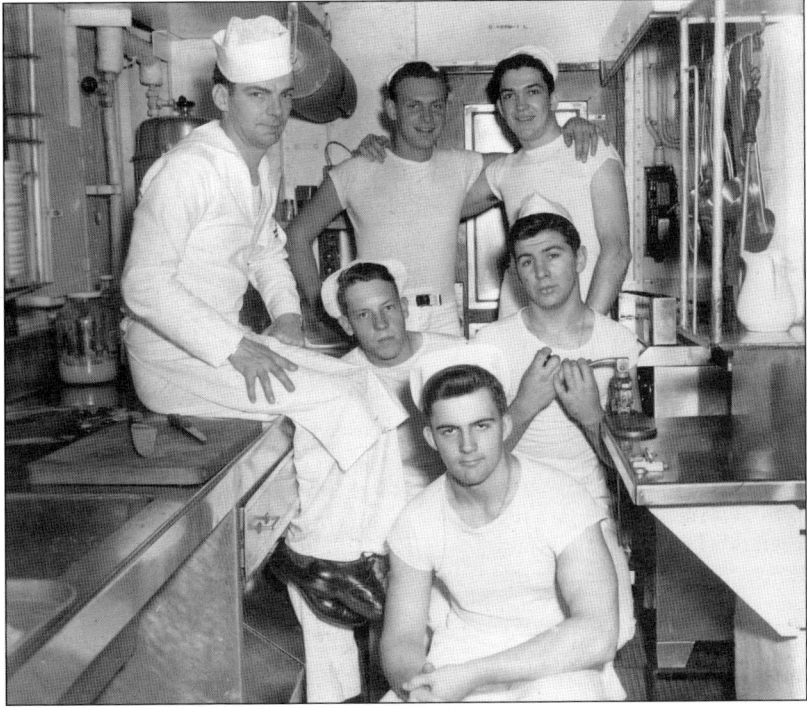

Chief petty officer (CPO) is the highest-ranking noncommissioned rank in the Navy. This photograph shows *Alabama*'s CPO pantry and crew. Breakfast was cooked in the pantry, but other CPO meals were prepared in the main crew galley and then brought to the pantry to be prepped and served to more than 50 CPOs.

Above, several of *Alabama*'s chief petty officers enjoy a steak meal served on plates with side dishes. Below, several new CPOs eat their first meal in the CPO dining area. They were expecting steaks like the ones served to other CPOs, but instead, they were served an induction meal of beans served in miniature pig troughs, with oversized galley spoons as their utensils. Note the open can of beans on the table.

Dining stations were set up around *Alabama*'s main galley wherever space was available, including on the outside wall of turret number three. Raised edges helped secure items in rough seas. The main mess areas used metal serving trays and china cups and bowls.

Good cooks are always appreciated in the military. Many generals and admirals had personal cooks, or chefs for their personal use and for their staff. Dillis Racho, a favorite chef during his time aboard *Alabama* while serving Rear Admiral Hanson, is seen below preparing a lobster. Racho was known for his culinary skills and for passing along tips to other cooks.

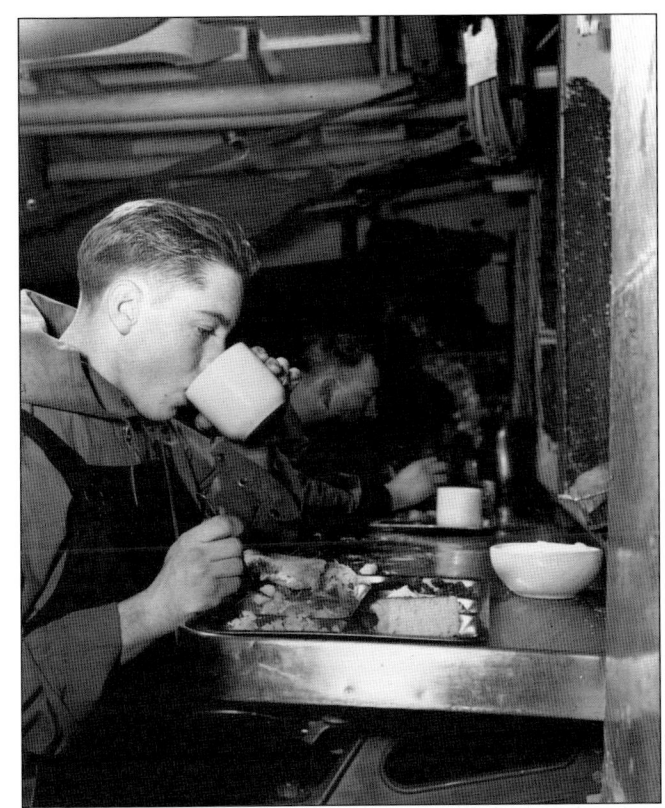

Robert William Andrew "Bob" Feller, the famous Cleveland Indians pitcher, joined the Navy two days after the bombing of Pearl Harbor. CPO Feller is seen here taking a turn at bat in a baseball game in the Pacific while serving on the *Alabama*. Teams were formed on larger ships and bases. They would play each other when possible in organized events, or just for pickup games. *Alabama*'s team was the fleet champion.

Exercise was important while at sea. Various activities ranged from simple modified volleyball games—using a variety of items besides a ball, which could bounce overboard—to organized sports, including wrestling and boxing. This picture from the *Alabama* archives shows a modified volleyball game taking place on a ship, possibly not the *Alabama*, in the Pacific.

Boxing was a popular sport during the war. Fighters representing other ships and militaries would fight during large matches. Smaller events pit boxers from the same ship against each other. The fights provided much-needed entertainment, as well as bragging rights for the winner. These photographs show numerous spectators aboard *Alabama*. They watched from chairs, benches, and the turrets, which were turned towards the sides of the ship to open up extra space. *Alabama* fielded 15 to 16 boxers on her team.

Movies were always popular whenever they were available. Efforts were made to provide popular movies to troops in every service branch. Popular titles were often traded between ships. The original caption for this photograph, from the USS *Alabama* war diary, reads "Carpenter gets ready."

Here, a movie screen has been set up on deck and the ship's band is ready to perform during one of *Alabama*'s movie nights. Movies were shown on deck every night if the ship was not in a combat area and the weather permitted. In other cases, the movies were shown in the crew's messing area and wardroom.

The USS *Alabama* had its own library, which handled all incoming and outgoing mail. Reading mail and sharing events from loved ones at home (right) was a favorite pastime on all fronts of the war. The photograph below shows mail being censored to ensure that no sensitive information was improperly released.

Members of *Alabama*'s crew take advantage of some downtime on a hot day. This photograph, from *Alabama*'s war diary, is captioned "Swimming Hole." Safety swimmers are manning the small rafts.

Several of *Alabama*'s crew enjoy a beach party in the Pacific. According to *Alabama*'s war diary, this photograph was taken in January 1944 on Enewetak Atoll. *Alabama* served within the fast carrier task force and was assigned to task force 58.2. She participated in the taking of several islands and atolls around the Marshall Islands.

These two photographs show the celebration that took place when *Alabama* crossed the equator. Sailors marked the occasion by having a celebration and by inducting first-time "line-crossers" into the fold. These photographs show the temporary pool that was set up under turret three for inductees to take a "deep six" plunge.

These two ceremony photographs show one of the first-time line-crossers taking a few minutes after his "deep six" swim with members of "King Neptune's court." The celebrations included music, games, the deep six swim, and other activities, including running the gauntlet, in which the first-time crossers ran the gauntlet across *Alabama*'s deck as they were pelted by wet clothes and hosed down with water.

Six
Faith and Healing at Sea

Just because the men were at sea did not mean that they could not go to church or have a toothache treated. This was at least the case for Navy personnel on larger ships serving in World War II. Those on board smaller vessels may have had to wait until one of the bigger craft pulled alongside before they could receive the level of care they needed.

It was not that the Navy thought smaller vessels in the fleet did not need these services; it was simply a matter of space and logistics. A smaller ship with a defined purpose did not have enough square footage to house the health services that a larger craft such as a battleship or aircraft carrier could.

The plan was that a battleship or carrier with specialized services could cater to the needs of smaller craft in a task force or fleet. That principal holds true today. If an injury happened aboard a World War II–era ship that the medical staff did not have the facilities to properly treat, then every effort was made to transport the patient to a ship that could provide the care needed.

The same thought process applies to the spiritual needs of the crew on a smaller vessel. The only difference was that sailors were not transferred to a larger vessel for that purpose; instead, chaplains from larger ships were sent to smaller vessels in order to provide services. The chaplain would often be "rigged" into a basket and transferred between ships while both vessels were underway at sea.

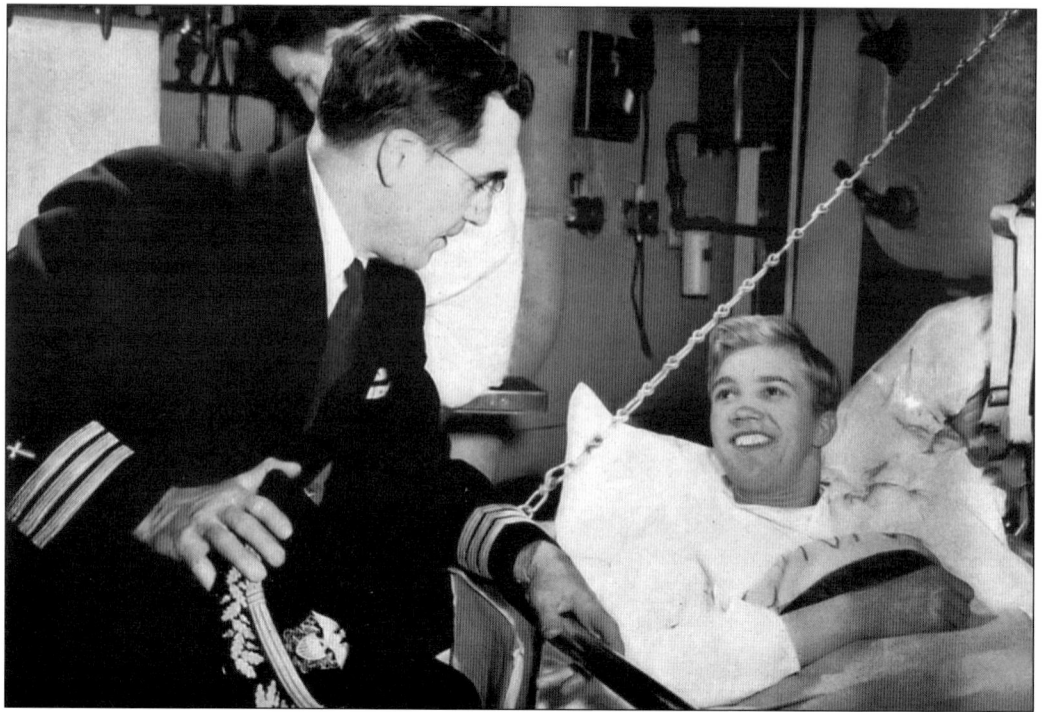

Presbyterian chaplain George L. Markle spends a few minutes with an *Alabama* crew member identified as Nolin while visiting sick bay. *Alabama* was manned by three chaplains during the course of her service. Markle was the first chaplain on board, serving from August 10, 1942, to February 10, 1943.

Church services were held on *Alabama*'s deck when the weather allowed. This service was held under a tarp, indicating that it may have been a special service such as communion. Another possibility, indicated by the small crowd, was that the tarp was used simply to protect those present from the heat of the day. Tarps could be rigged rather quickly at the request of the chaplain.

These two photographs show services on the deck of *Alabama* taking advantage of the space aft near turret three. This was the same area that hosted movies and special events such as boxing matches. During bad weather, services were held below deck, wherever space allowed. Most often, they took place in the main crew messing area.

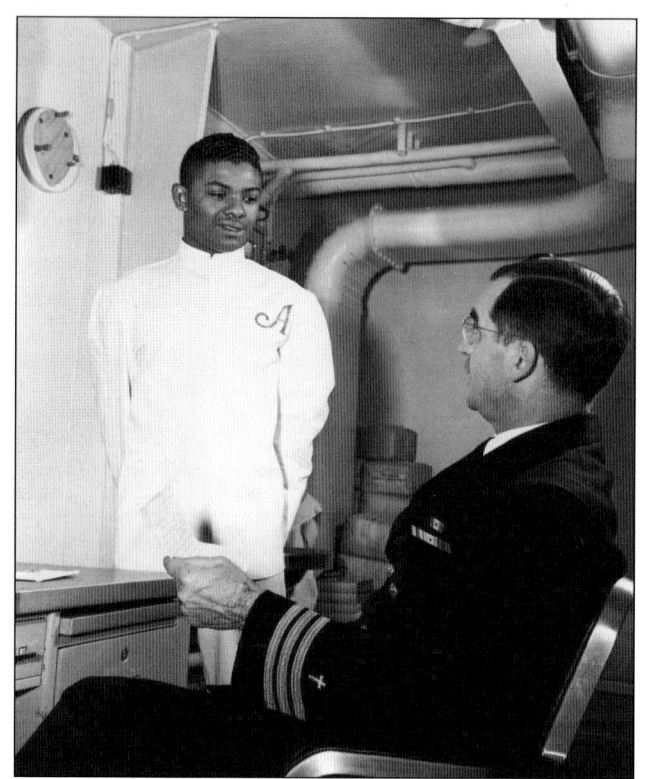

Chaplain George L. Markle is seen in these two photographs with individual crew members aboard *Alabama*. He is seen at left with one of *Alabama*'s African American crew members. The photograph below was taken in the ship's reading room, which housed a wide variety of books, including Bibles. When the war started, African Americans mainly served as mess men, attendants, porters, and cooks, but they were still assigned to combat battle stations. Later in the war, in 1944, the destroyer escort USS *Mason* was commissioned with a crew of 150 African American sailors and petty officers.

Alabama's other two chaplains are seen here. Episcopal chaplain Charles L. Glenn (left) served from October 30, 1944, to October 15, 1945. Roman Catholic chaplain Edward Xavier Praino (right) served on Alabama from October 25, 1943, to February 20, 1945.

This photograph shows a burial at sea onboard Alabama. No crew members were killed due to enemy fire while on Alabama. However, there were lives lost. Five crew members were killed in an accident during air attacks in the Pacific, when the number nine gun mount accidentally fired into the number five mount. Five men were killed and another eleven were injured. One of the crewmen killed was African American.

The only pennant allowed to fly over the American flag was the church pennant, which indicated that church services were taking place aboard *Alabama*.

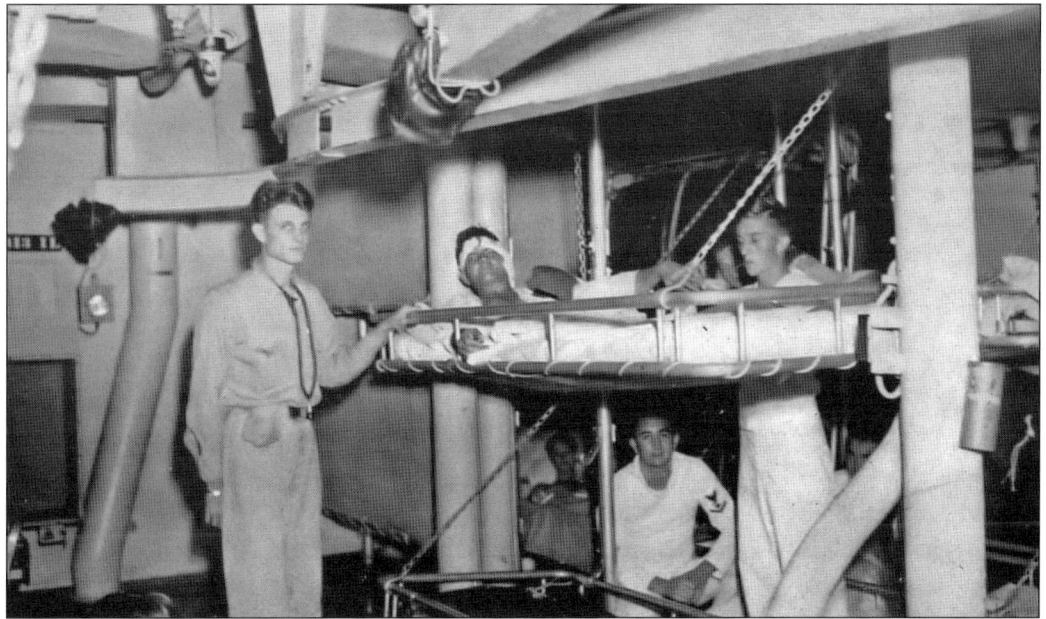

The USS *Alabama* had impressive medical facilities on board. The area commonly referred to as the sick bay actually consisted of an array of rooms, including traditional hospital areas and labs, an eye, ear, nose, and throat office, a dispensary, and a dentist's office.

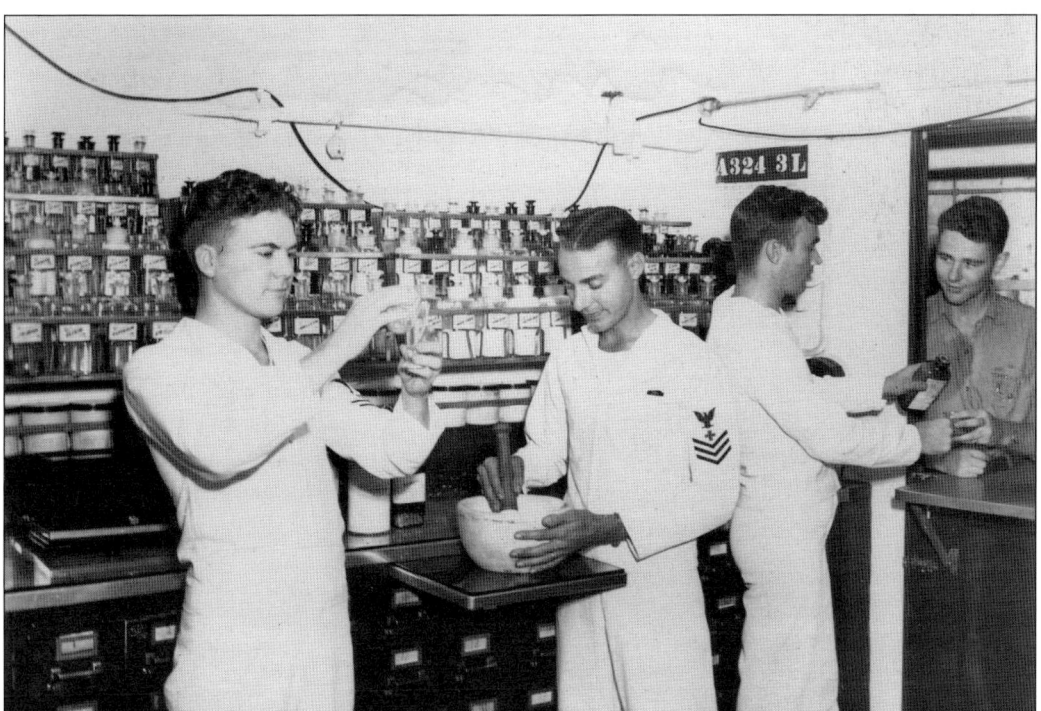

This photograph shows crew members working in the dispensary on board *Alabama*. The dispensary filled prescriptions for crew members, as would be done in any normal pharmacy. Prescriptions were written by the ship's doctors.

Ships that lacked the proper facilities needed to handle a severe injury often shipped their patients to the *Alabama* for treatment. This photograph shows one of the most common ways to get injured men from one ship to the *Alabama* while underway.

Emergency and basic first-aid training were vital to a ship's crew. The goal of the training was to make sure that everyone could properly care for an injured fellow crew member and transport them safely to the sick bay. In this way, every member of the crew became part of the medical staff. Here, an *Alabama* crew member is secured during training.

This photograph shows one of *Alabama*'s two dentist's chairs. As with the other doctors serving on *Alabama*, the dentists saw patients from other ships as well. This photograph shows the renovated office. The light fog at the top left and bottom of the photograph is from light reflecting off the Plexiglas in the doorway. (Courtesy of Kent Whitaker.)

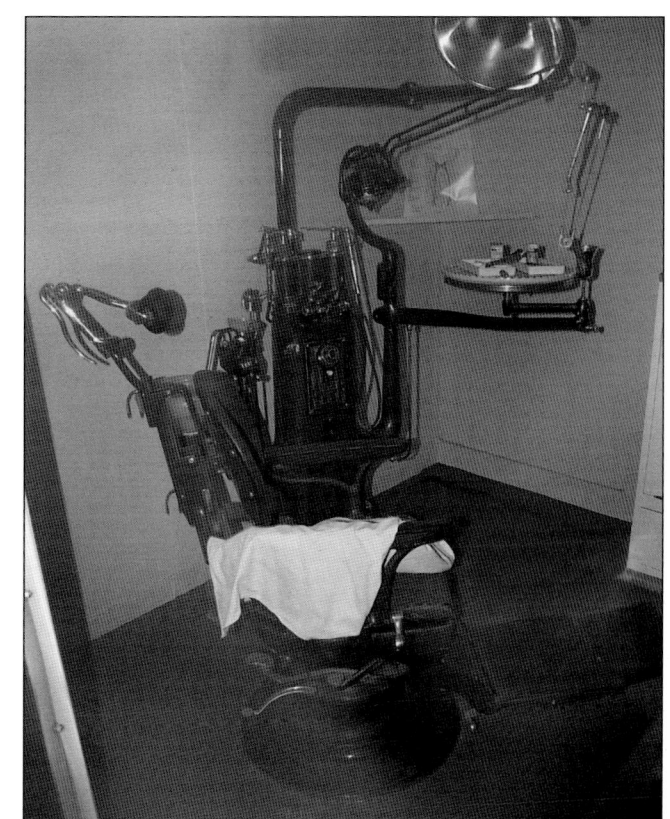

Alabama's medical staff is seen below taking an X-ray of a sailor's injured lower leg. Medical equipment such as this was relegated to larger vessels such as the *Alabama* and other battleships and carriers. Some hospital ships and cruise liners that were converted to transport ships may have had similar equipment.

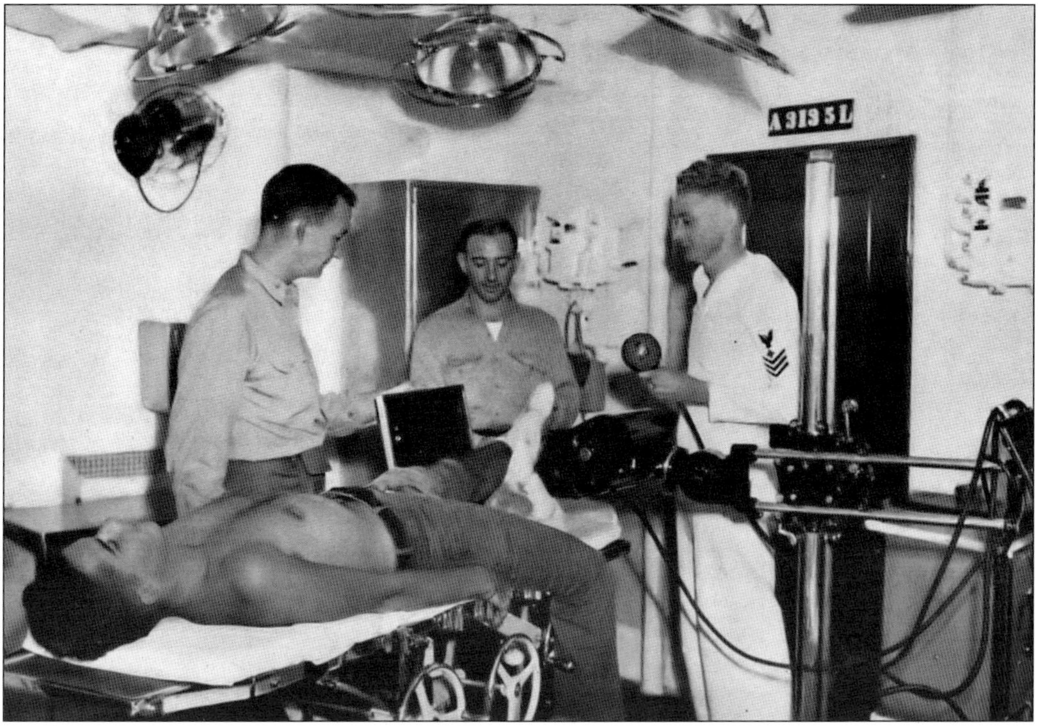

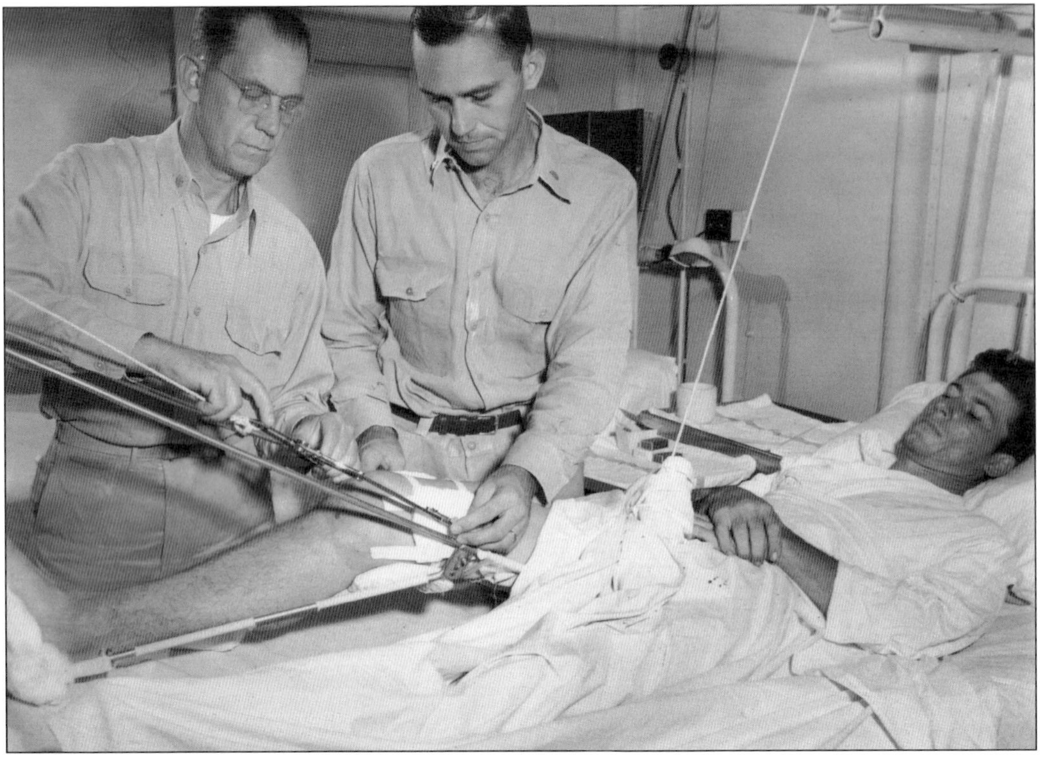

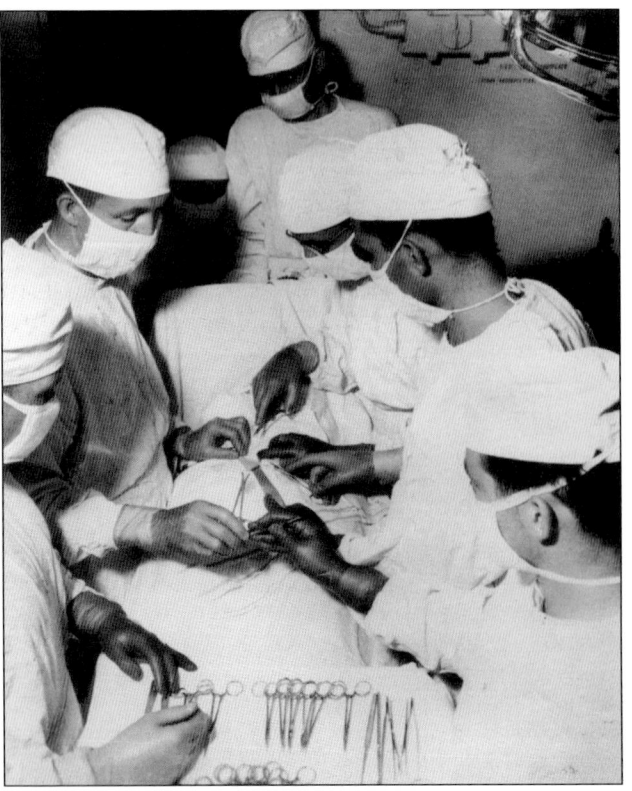

USS *Stembel* seaman Walter Metz (above), along with six other sailors, suffered injuries during a storm in July 1945. All seven were sent to the USS *Alabama* for treatment. Metz's leg was broken in two places, which required it to be weighted, braced, and set with a wire drilled through the bone. His shipmates were treated and eventually sent back to *Stembel*, but Metz finished the war onboard *Alabama*.

This photograph shows *Alabama*'s medical staff during surgery. Not many ships in the Allied forces had full operating rooms. Typical injuries aboard vessels such as a battleship were hand and foot injuries caused by falls or crush accidents. Stitches for cuts caused by banging one's head against a low bulkhead door were common.

Seven

EARNING NINE BATTLE STARS

The USS *Alabama* (BB-60) earned nine battle stars during World War II. These honors did not start when the "Mighty A" entered into her first battle; the effort to earn these wartime honors began when her crew first took her to open waters for sea trials.

Alabama served in both the Atlantic and Pacific theaters. She was present when convoys needed to be protected early in the war and when offensive task forces needed protection during the war and towards its end. Her crew was present when she led the Allied fleet into Tokyo Bay and when servicemen simply needed a ride home to the states.

Alabama's crew earned battle stars for their service in 1943, 1944, and 1945. The first was for action in the Gilbert Islands from November 19 to December 8, 1943. The next six battle stars came in 1944 for action in the Marshall Island operations, the Hollandia operations, the Marianas operations, the western Caroline Islands operations, and the Leyte operation. All of those operations pushed the Japanese forces off occupied islands.

In 1945, *Alabama* added two more battle stars for her service during operations leading to the eventual assaults on mainland Japan. These were the Okinawa Gunto operations and the Third Pacific Fleet operations against Japan.

Alabama is credited with shooting down 22 enemy aircraft, providing cover and bombardments for key invasions during the war, and for sounding the alarm to an entire task force that an enemy attack was nearing. This alert came during a critical point in the war and was due to the fact that the *Alabama*'s crew took it upon themselves to install the newest radar system while underway at sea.

The USS *Alabama* is seen here in March 1943 while underway in the Atlantic. This photograph shows the wide range that her "Big Guns" could cover. Turret two is rotated almost completely aft. Each turret could rotate up to 270 degrees.

Alabama is seen here on August 20, 1943, as she steams towards Norfolk, Virginia, shortly after being refit and after serving in the Atlantic with the British Home Fleet.

Alabama is seen here on her shakedown cruise prior to heading towards the North Atlantic. It was during her early shakedown cruise that *Alabama* ran aground on an unmarked sandbar between the Chesapeake Bay and Baltimore. She is painted in her Measure 12 (modified) camouflage paint scheme. This photograph was taken in Casco Bay, Maine, in December 1942.

In the foreground of this photograph is the USS *South Dakota* (BB-57), with the *Alabama* alongside. The photograph was taken from the deck of HMS *King George V* when *Alabama* was serving in the Atlantic with the British Home Fleet in 1943.

The *Alabama*'s superstructure is seen here in December 1942. Mounted atop the superstructure is the fire control radar for her "Big Guns." The second radar, below the flags to the right, is the fire control radar for the secondary battery. The smaller loop-shaped piece of equipment to the lower left is the radio direction finder.

Below, in the North Atlantic, members of *Alabama*'s crew spend a few minutes with Royal Marines. The British moved much of their naval forces to the Mediterranean theater prior to the invasion of Sicily. *Alabama* and *South Dakota* were assigned to the Royal Navy Home Fleet to help protect northern convoy routes.

Alabama was home to three OS2U aircraft, called Kingfishers. The planes' main role was to observe the accuracy of shots fired from *Alabama*'s 16-inch main battery. The planes also served in several other roles, such as being a taxi and for search and rescue missions.

Alabama's Kingfisher aircraft were launched off catapults located on both the port and starboard side of *Alabama*'s aft deck. This launch took place on the starboard catapult. The planes would return to *Alabama* by landing at sea. *Alabama*'s crane would then lift them back onto the deck.

The *Alabama* did not have a hangar for her aircraft, so they were exposed to the elements unless covered by tarps in extreme conditions. This meant that inspections and routine maintenance of the planes was done on a constant basis. The crane used to lift and move the planes is in the background.

This photograph shows one of the targets in the Marshall Islands being bombarded by *Alabama*'s 16-inch "Big Guns." The caption from *Alabama*'s war diary simply says, "The Target Burns." The Kingfisher pilots spotted the targets and communicated accuracy back to the *Alabama*. An airstrip on the island is in the foreground.

Raphael Semmes Jr. was the great-grandson of Capt. Raphael Semmes, who commanded CSS *Alabama* during the Civil War. Semmes Jr., seen here in 1960, served on the USS *Alabama* (BB-60) as a lieutenant from November 1942 to July 1943, during *Alabama*'s North Atlantic convoy escort duties with the British Home Fleet and during the Murmansk Run. He was the senior aviator and the head of the V (Aviation) Division.

While this photograph appears to show *Alabama* taking on nearby enemy fire, the plumes on the horizon are actually practice shots fired from another battleship. In order to keep in training, the crews of the "Big Guns" on each battleship would target behind another ship for practice. Pilots also used the larger ships for practice bombing and torpedo runs.

Crew members aboard *Alabama* seem happy about the Allied air cover overhead. This undated photograph from the USS *Alabama* archives was taken while she was serving in the Pacific. Note that none of the sailors are wearing helmets, indicating that there was no immediate enemy threat.

This photograph, taken from *Alabama*, shows a large splash caused by a Japanese plane crashing into the Pacific Ocean after it was shot down. Two of *Alabama*'s 20-millimeter single antiaircraft machine guns are seen in action.

This photograph was taken from the deck of the USS *Essex* (CV-9) on October 14, 1944. It shows splashes on the water's surface caused by the impact of a torpedo just released from a Japanese torpedo bomber. Other splashes are from gunners shooting at the bomber and the torpedo. *Alabama* is seen on the horizon to the right.

Air protection is flying above *Alabama* in this photograph. While serving in the Pacific, *Alabama* and her crew were most often part of carrier groups or task forces, which combined multiple types of ships of all sizes. The *Alabama* war diary refers to air support as being "Guardian Angels."

These two photographs of the USS *Enterprise* were taken during the assault in Okinawa on March 20, 1945. They both show the aftermath just seconds after a Japanese plane crashed into her. Both photographs show *Enterprise* as a smoke billow rises above her. The photograph below was taken by one of the combat photographers aboard *Alabama*.

American planes are seen here leaving on a mission. The USS *Alabama* patrolled areas around the Marianas and the South Pacific in 1944 and 1945 in order to protect the American landing forces and carriers. On October 6, 1944, *Alabama* sailed with Task Force 38 to aid in the liberation of the Philippines and then the assaults on Okinawa.

Alabama supported aircraft carriers during air raids on the islands of Tinian, Saipan, and Guam in June 1944. This photograph was taken as a splash from a Japanese bomb lands—a near miss just abreast of *Alabama*'s turret two.

This photograph was taken in the Pacific during assaults on several Japanese-held islands, including the assault against Okinawa. The main battery of 16-inch guns are silent as the antiaircraft guns do their jobs.

This photograph was taken in May 1945 while *Alabama* was serving during assaults on Okinawa. It shows Japanese bombers as they attempt suicide attacks against carriers in Task Force 58.

This photograph was taken aboard the USS *South Dakota* (BB-57). It shows battleships USS *Indiana* (BB-58), USS *Massachusetts* (BB-59), and USS *Alabama* (BB-60) forming a line as they steamed through the Pacific towards the war's end in 1945.

This photograph was taken in Seattle, Washington, on Victory Over Japan Day. The photograph is dated August 14, 1945, as word trickled back to the states that Japan was surrendering. The photograph, originally from the archives of the Museum of History and Industry, is now in the archives of the USS *Alabama*. The original caption identifies the sailor on the light post as USS *Alabama* crewman Orlando F. Scarpelli.

These two photographs show USS *Alabama* at the end of the war, in September 1945. She is seen above leaving Pearl Harbor on September 10, as she steams to the West Coast. She is seen below nearing San Francisco. Her decks are packed with her regular crew as well as the numerous servicemen who hitched a ride home.

Tugboats were a vital part of any wartime harbor, as many ships were too large to safely navigate in shallow waters. Experienced tug crews familiar with local waters could gently guide and nudge ships of all sizes. After *Alabama* arrived home, she spent several weeks ferrying troops to ports along the West Coast.

As the USS *Alabama* arrived home, her crew was welcomed back by an appreciative nation. This photograph was taken during a ship's dance for the crew of *Alabama* at the end of the war, at the Lakewood Country Club in Los Angeles.

Alabama is nudged into port by a tugboat at Pier 90 in the Puget Sound Naval Shipyard in September 1946. Moored behind her is the carrier USS *Bunker Hill* (CV-17). The superstructure of another battleship can be seen above the storage and pier buildings to the left.

This photograph was taken from Telegraph Hill in San Francisco after the war's end. Three of the four South Dakota–class battleships wait to be decommissioned, as their missions have ended. They are, from left to right, USS *Indiana* (BB-58), USS *Massachusetts* (BB-59), and USS *Alabama* (BB-60).

Eight
Bringing Alabama Home

After World War II, USS *Alabama* (BB-60) was placed in the reserve fleet of the Navy. This designation usually meant that a ship would be held until she was called to life once again or deemed unsuitable for service.

This is where the ship known as both the "Mighty A" and the "Lucky A" began her journey home to her namesake state. In May 1962, a short newspaper article stating that the USS *Alabama* was being scheduled for scrapping by the Navy spurred a wave of support from people across the state.

At the request of Gov. George B. Wallace, the Navy delayed its scrapping plans long enough to see if the state could put together a plan to move the ship to a new home. The delay proved to be successful. A committee was formed to organize the efforts, and soon, businesses, citizens, and schoolchildren across the state of Alabama raised enough money to bring the famed battleship to Alabama.

Alabama was saved from the scrap pile with the help of $100,000 raised by school children across the state. It came mostly in the form of pocket change. In total, almost $1 million was raised across the state. The process of preparing the ship for her long tow home began. The 5,600-mile tow from the reserve fleet shipyards near Seattle to her new home in Mobile set the record for the longest nonmilitary tow in history.

Today, the USS *Alabama* still serves majestically, as a learning museum ship at Battleship Memorial Park in Mobile Bay. She is joined by the USS *Drum*, a World War II submarine, as well as a variety of aircraft and military vehicles. All stand in honor of all who have served the United States of America.

On June 16, 1964, Capt. Jim Thwing of the Navy Reserves received the decommissioned battleship *Alabama* from the Navy on behalf of Governor Wallace and the people of Alabama. Captain Thwing served as the state's West Coast representative during negotiations and then spent seven weeks in Mobile during the ship's renovations.

Herman Hardell of Port Orchard, Washington, worked as a leaderman rigger at the Puget Sound Naval Shipyard. He is seen below watching *Alabama* leave her moorings for a short trip to the Todd Shipyard in Seattle, where she was rigged for the longest nonmilitary tow in history. Hardwell recalled that he had taken care of "six battlewagons" over his then-24-year career.

The battleship known as the "Mighty A" and the "Lucky A" is seen in these two photographs as tugs move her from the Navy shipyard in Bremerton, Washington, to the Todd Shipyard in Seattle, and, eventually, towards the open sea. Soon after, she began her record-breaking tow of 5,600 miles to Mobile. The trip took her down the West Coast, through the Panama Canal, into the Caribbean, and across the Gulf of Mexico.

Alabama is seen here in Seattle on July 7, 1964, as she is ceremoniously transferred to the state of Alabama. From left to right are Albert M. Beldsoe, "King Neptune" for the Seattle festival called Seafair; Stephens G. Croom, secretary of the Battleship Alabama Commission; Arlene Hinderlie, "Queen of Seafair;" and Charles L. McLafferty of Mobile, who served aboard *Alabama*.

Funding was needed to bring the battleship from Bremerton, Washington to Mobile. Schoolchildren across the state donated more than $100,000, mostly in the form of pocket change and lunch money. For a donation, the children received free admission cards. A corporate fundraising effort was also put into place, which raised almost $1 million.

Alabama is seen here underway as it headed towards Mobile. The city of Seattle is in the background, with the Space Needle visible. Also visible are *Alabama*'s propellers, which were removed for the journey. They are secured to the forward deck, above the number 60 on the hull.

Alabama is seen here in July 1964 as her tugs guide her through the first portion of the Panama Canal. She is just about to pass under the Thatcher Ferry Bridge. In the background, portions of the Rodman Naval Station can be seen. The station was handed over to the government of Panama in 1999.

This July 1964 aerial photograph shows the narrow width of the locks along the Panama Canal. The canal is a combination of open water and locks, which raise and lower ships through the mountains and valleys from the Pacific Ocean to the Atlantic.

Alabama and her final crew are seen here. During her tow home, she receives a salute from active-duty Navy personnel stationed in Panama. This July 1964 photograph indicates that it was taken as *Alabama* made her way out of the Miraflores Locks, the first of three locks in the canal transit.

The "Mighty A" draws a crowd during her transit through the canal, as people take the opportunity to see her pass by. This photograph shows *Alabama* making her way through a channel, with one of the canal locks in the background.

The USS *Lexington*, which was still in service after World War II, sailed with *Alabama* for a short time in the Gulf of Mexico during her tow. In June 1944, *Alabama* was protecting the carriers of Task Force 58, in which *Lexington* was assigned. *Alabama*'s new SK-2 radar system, which had been installed at sea by the crew without the use of shipyard cranes, provided Task Force 58 with an early warning of approaching enemy craft, giving the *Lexington* and the other carriers time to launch their planes and for the task force to ready defenses. The lopsided Allied victory became known as the "Great Marianas Turkey Shoot."

World's Longest Tow

This postcard from the Alabama Post Card Company shows the tug *Sea Ranger*, one of the three primary seagoing tugs that guided *Alabama* along her 5,600-mile journey. The second tug was the 110-foot-long *Sea Robin*. Both had a 12-man crew. While underway, the *Sea Robin* was replaced by the third tug, *Sea Lion*.

The tugboats worked hard during what turned out to be a dangerous voyage. A malfunction in *Sea Lion*'s steering gear caused the vessel to sink en route to the Panama Canal. The incident took the life of Ira H. Goltry, a Merchant Marine captain and chief mate of the *Sea Lion*. After the sinking, the Mobile-based tug *Margaret Walsh* was sent to take *Sea Lion*'s place.

This aerial photograph shows *Alabama* being towed as she nears Mobile. Several small boats are now sailing alongside, and what appears to be a small Coast Guard cutter trails behind her. The towlines that the tugs used were steel hawsers (rope). Each was 2.5 inches thick and 2,800 feet long.

This postcard from the Alabama Post Card Company shows the growing number of boats that joined Alabama and her tugs as she traveled across Mobile Bay.

These two photographs were taken by Jack Gibson Photography of Mobile and are from damaged contact sheets in the USS Alabama archives. Both show the Alabama making her way through Mobile Bay from the perspective of people in the small armada of local pleasure boats that escorted her across the bay.

Robert S. Edington, an Alabama state senator (far right) is shown with a group of unidentified people as Alabama comes home. Edington wrote the enabling legislation passed by the Alabama legislature creating the USS Alabama Battleship Commission, an agency of the state of Alabama, in 1963. Without the commission, the battleship probably would never have been given to the state. Edington served as chairman of the commission and is still on the commission as of 2013.

These two photographs, taken by Thigpen Photography of Mobile, show the work progression on Battleship Memorial Park in 1964. A total of 2.9 million cubic yards of bay bottom were dredged to make a channel in order to bring *Alabama* in. This material was then used to make the first 75 acres of the park. The photograph below shows the expanding landmass and the *Alabama* in place.

The uncredited photograph above shows *Alabama* in place as work is still being done to fill in land for Battleship Memorial Park. Despite the graininess and damage to the photograph, one can still make out some of the equipment being used as fill material, which was brought in from the bay. The photograph below, by Thigpen Photography of Mobile, shows the completed Battleship Memorial Park in 1965. The battleship, the main parking area, parked cars, and the ship's store can all be seen.

Workers are seen here using heavy equipment to build the foundation for Battleship Memorial Park. Much of the material and fill used was dredged from Mobile Bay. To the left, the black supports are in place for one of two access ramps being built.

These workers man a pile driver used to build docks. The machine made it possible to properly shore up the edges of the park and maintain the stability of the fill material. While *Alabama* appears to be floating, she is actually set over 20 feet into the bottom of Mobile Bay. In the background, the second access ramp can be seen.

Workers hang overboard and work from floating platforms in order to clean and repair the outside of *Alabama* to properly restore her. Similar work was underway on the interior of the ship as well.

The photograph below shows some of the work done at the Todd Shipyard in Seattle to prepare *Alabama* for her tow. *Alabama*'s four propellers, or "screws," were removed and secured to the deck for the trip. Here, a propeller is being moved to its display position on the deck. It was displayed there until 1992. Two of the propellers were sold and two are still on display.

123

Work is underway on *Alabama*'s renovation in 1964. Many renovations were done prior to the ship being opened to the public, but renovations and repairs are constantly being done to this day. The main renovations done in the first few years, and those done before opening day, involved making sure the ship was safe for guests.

Below, work progresses on the deck of *Alabama*, moored at her new home in Mobile. This photograph also shows restoration work being done on turret three. Noticeably missing is the 16-inch "Big Gun" protective skirting, known as bloomers.

On January 9, 1965, the battleship celebrated her grand opening. More than 5,000 people visited the *Alabama* that first day. Among the first visitors were "*Alabama* Rear Admirals," guests who contributed $100 or more. Also present were families, schoolchildren, politicians, and veterans of World War II. The photograph above, by John Spottswood Photography of Mobile, shows one of the younger guests posing on shore and is captioned "Dedication Boy." At right, a youngster tries one of *Alabama*'s antiaircraft machine guns.

One of Alabama's four propellers is on display outside of the ship's store. It is one of the first ship exhibits that visitors had a chance to view. The propellers, often called "screws," powered Alabama at a speed of 28 knots and were made of bronze. The propeller seen here weighs 18.5 tons and is 17 feet and 4.5 inches across. (Courtesy of Kent Whitaker.)

Alabama is seen here with some of the static military displays located at Battleship Memorial Park. Today, the park is also home to the World War II submarine USS Drum. The air museum hangar features more than 20 different aircraft, including a Blackbird spy plane. Memorials, boats, planes, and monuments to servicemen and women of all eras and service branches are located on the park grounds. (Courtesy of Kent Whitaker.)

About the USS Alabama Battleship Memorial Park

USS Alabama Battleship Memorial Park is located in Mobile Bay. The park is home to the battleship USS *Alabama*, the submarine USS *Drum*, an aircraft pavilion, and many outdoor static displays featuring aircraft, tanks, and equipment from all service branches. Also located on the park grounds are memorials for those who served in World War II, the Korean War, and the Vietnam War, as well as a memorial for POW-MIA.

The park hosts special events, reunions, military dinners, and stays for organized youth groups. Located on the park grounds is the ship's store, which features a wide range of gift items, books, and a restaurant. The battleship and park are funded by ticket sales and donations.

Battleship Memorial Park is open every day except Christmas, from 8:00 a.m. to 6:00 p.m. April through September and from 8:00 a.m. to 5:00 p.m. October through March. Please note that the last ticket is sold one hour prior to closing.

Telephone: (251) 433-2703
Address: 2703 Battleship Parkway, Mobile, AL 36602
www.ussalabama.com

Discover Thousands of Local History Books
Featuring Millions of Vintage Images

Arcadia Publishing, the leading local history publisher in the United States, is committed to making history accessible and meaningful through publishing books that celebrate and preserve the heritage of America's people and places.

Find more books like this at
www.arcadiapublishing.com

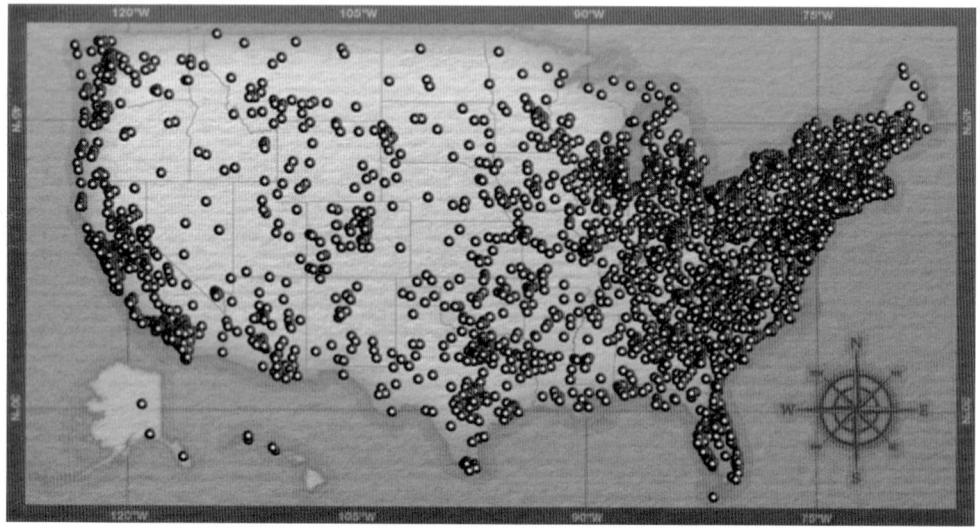

Search for your hometown history, your old stomping grounds, and even your favorite sports team.

Consistent with our mission to preserve history on a local level, this book was printed in South Carolina on American-made paper and manufactured entirely in the United States. Products carrying the accredited Forest Stewardship Council (FSC) label are printed on 100 percent FSC-certified paper.